Leisure and Tourism Landscapes

T0187833

The theme of this important new book is the influence of leisure and tourism on the production, representation and consumption of landscape. Increasingly significant as mediators of spatial identity and meaning, leisure, tourism, culture and heritage are only now beginning to be located within the rapidly evolving discourses of poststructuralist geographies. Embracing poststructuralist theory to develop analyses of space, place and landscape in everyday life, this book makes a vital contribution to these discourses.

The first half of the book focuses on different 'ways of seeing' or representing landscape. The second half of the book examines different forms of productive consumption in leisure and tourism. Both symbolic and material spaces of leisure and tourism are examined in relation to urban and rural landscapes, heritage landscapes, gendered landscapes and landscapes of sexuality and desire.

Leisure and Tourism Landscapes draws on a range of interdisciplinary subject areas, such as leisure and tourism studies, cultural studies and gender studies, to supplement analyses from geography, sociology, history and literature. It promises to redefine the way we view our surroundings.

Cara Aitchison is Reader in Leisure Policy and Cultural Theory and Head of the Leisure and Sport Research Unit at Cheltenham and Gloucester College of Higher Education. **Nicola E. Macleod** is Senior Lecturer in Leisure and Cultural Management in the Business School at the University of North London. **Stephen J. Shaw** is Senior Lecturer in Transport and Tourism Management in the Business School at the University of North London.

Routledge Advances in Tourism
Series Editors: Brian Goodall and Gregory Ashworth

Leisure and Tourism Landscapes

Social and cultural geographies

Cara Aitchison, Nicola E. MacLeod and Stephen J. Shaw

London and New York

First published 2000
by Routledge
11 New Fetter Lane, London EC4P 4EE

Simultaneously published in the USA and Canada
by Routledge
29 West 35th Street, New York, NY 10001

Reprinted 2001

First published in paperback 2002

Routledge is an imprint of the Taylor & Francis Group

© 2000, 2002 Cara Aitchison, Nicola E. MacLeod and
Stephen J. Shaw

British Library Cataloguing in Publication Data
A catalogue record for this book is available from the British
Library.

Library of Congress Cataloging in Publication Data
   Aitchison, Cara.
   Leisure and tourism landscapes: social and cultural geographies/
   Cara Aitchison, Nicola E. MacLeod and Stephen J. Shaw.
      p.   cm.
   Includes bibliographical references and index.
   1. Leisure—Social aspects. 2. Tourism—Social aspects. 3. Human
   geography. I. MacLeod, Nicola E., 1966– . II. Shaw, Stephen J.,
   1955– .
   III. Title.
   GV14.5. A57 2000
   306.4'812–dc21   00–030820

ISBN 0–415–17060–5 (Hbk)
ISBN 0–415–27166–5 (Pbk)

Contents

Acknowledgements

Cara Aitchison would like to thank her parents for introducing her to Scotland's varied material and symbolic landscapes, Carole for sharing journeys to landscapes further afield, and Cheltenham and Gloucester College of Higher Education for providing the time and space to theorise these landscapes of leisure and tourism.

Nicola E. MacLeod would like to acknowledge the support of friends, family and colleagues in the writing of *Leisure and Tourism Landscapes*. She dedicates the book to the memory of her mother, June MacLeod (née Chalmers) who, in pointing out 'scenery' on long car journeys, first encouraged an appreciation of the Scottish landscape.

Stephen J. Shaw would like to express deep gratitude to his partner Marguerite for her philosophy, patience and humour during the preparation of this manuscript.

Cara, Nicola and Stephen wish to acknowledge the dedicated support of the Routledge team, especially Simon Whitmore, now Associate Editor for Sport and Leisure Studies at Routledge, and Susan Leaper, Vicky Squires and Sue Edwards at Florence Production Limited.

The authors and the publishers would like to thank the following for permission to use copyright material.

Taylor & Francis for excerpts from: 'New cultural geographies: the spatiality of leisure, gender and sexuality', written by Cara Aitchison and first published in 1999 in *Leisure Studies* 18(1), 19–39; 'Poststructural feminist theories of representing Others: a response to the crisis in leisure studies' discourse', written by Cara Aitchison and first published in 2000 in *Leisure Studies* 19(3), 127–44; 'Heritage and nationalism: gender and the performance of power', written by Cara Aitchison and first published in 1999 in D. Crouch (ed.), *Leisure/Tourism Geographies: Practices and Geographical Knowledge*, pp. 59–73.

1 Introduction

A place for leisure and tourism?

This book is about the production, representation, consumption and (re)presentation of the British landscape. At first glance this appears to be a focus that has long captured the gaze of geographers. A more thorough viewing or reading, however, reveals that this gaze is redirected here through the lenses of leisure and tourism studies. Increasingly significant as mediators of spatial identity and meaning, leisure and tourism are only now beginning to be rendered visible, situated and *placed* within the rapidly evolving discourses of post-positivist or poststructuralist geographies.

Leisure and Tourism Landscapes: Social and Cultural Geographies attempts to embrace poststructuralist theory to engage with analyses of the construction of space, place and landscape in everyday life. Simultaneously, however, the book acknowledges the contribution of previous structuralist discourses to informing our understanding of landscape. As such, we draw on a range of interdisciplinary subject fields together with more traditional academic disciplines to explore the significance of symbolic and material spaces and places in the production, representation and consumption of leisure and tourism.

As S. Williams (1998: 172) asserts, 'The construction and subsequent consumption of tourist places is essentially a socio-cultural process'. This process has previously been conceived of as being mediated through a 'leisure industry', where commodification and consumption are controlled increasingly by professional and commercial interests (Adorno 1993). In contrast, writers such as de Certeau (1984) have stressed the importance of everyday life as a form of productive consumption where 'consumers', instead of being 'cultural dupes', are themselves constantly involved in remaking culture. This scope for agency allows us to see places as continually evolving landscapes with space for resistance, contestation, disruption and transgression of dominant discourses and wider hegemonic social and cultural relations. Whereas such subject fields as cultural studies have long given voice to 'subversive' narratives and discourses, it is only relatively

recently that geography has rendered audible those previously marginalised voices that have sought to construct a place for both people and
discourses excluded from traditional geographical agendas.

From geography to geographies?

In recent years geography has come to be seen as both a fractured and
contested discipline. 'What most of us too readily treat as a universal
discipline – a sort of "Geography-with-a-capital-G" – is really only one
sedimented and situated product of a series of intersecting historical geographies and colliding geographical experiments' (Barnes and Gregory 1997:
1). Indeed, the recently established journal *Tourism Geographies* explains
its adoption of the plural term 'geographies' as an attempt to:

> express a sensitivity and effort to reach out to the diversity of perspec-
> tives that fall under this subject matter, including both academic and
> applied research, regional traditions from Europe, North America,
> Asia-Pacific and other parts of the world, and disciplinary approaches
> from geographers and related professionals, such as anthropologists
> and other social scientists, landscape architects, urban and regional
> planners, and environmental scientists and managers.
>
> (*Tourism Geographies* 1999: n.p.)

This book attempts to embrace this plurality of geographies together
with the increasing academic freedom offered by the recent blurring of
disciplinary boundaries within the academy. Indeed, Haraway (1985: 191)
contends that postmodernism has developed 'as an argument for pleasure
in the confusion of boundaries and for responsibility in their construction'. This more flexible approach to engagement with a range of human
and social sciences has offered geographical insights from a range of disciplines and subject fields addressing leisure-related issues. These include:
sociology (Urry 1990, 1995), cultural studies (Bell 1991; Bell and Valentine
1997; Skelton and Valentine 1997), gender studies (Bell and Valentine
1995; Bondi 1992a, 1992b, 1998; Duncan 1996; Rose 1993, 1996;
Valentine 1996) and tourism studies (Crouch 1999; Edensor and Kothari
1996; Rojek and Urry 1997; Selwyn 1996; Urry 1990, 1995).

Within these disciplines and subject fields, questions relating to the interrelation of production, representation, consumption and (re)presentation
have recently been foregrounded. For example, in social and cultural
geographies we are aware of the iconography of landscape (Daniels and
Cosgrove 1988) the mapping of desire (Bell and Valentine 1995), the theorisation of 'BodySpace' (Duncan 1996), images of the street as refracted
through leisure and tourism (Fyfe 1998) and analyses of the interrelation
between shopping, place and identity (Miller, Jackson, Thrift, Holbrook
and Rowlands 1998). In sociology, we have witnessed increasing interest

in food, health and identity (Caplan 1997), body cultures (Bale and Philo 1998) and touring cultures (Rojek and Urry 1997), and in gender studies and cultural studies we are aware of the significance of Barbie culture (Rogers 1998), Nike culture (Goldman and Papson 1998) and internet culture (Porter 1997). In all of these disciplines and subject fields, leisure, culture, sport and tourism have been theorised as central sites and processes of identity construction, performance, contestation and negotiation. *Leisure and Tourism Landscapes* attempts to contextualise, illustrate and theorise these struggles for space. In doing so, we seek to identify and build on interdisciplinary synergies that can contribute to the ongoing development of social and cultural geographies.

Theorising the social–cultural nexus

Geography's cultural turn has encouraged analyses that take account of both structural and cultural determinants of spatiality. Simultaneously, these geographical interpretations of the social–cultural nexus have begun to address issues of leisure space, place and landscape. In their socio-cultural analyses contemporary social and cultural geographers have embraced Cooper's (1994) assertion that power is 'productive, relational and everywhere'. This emphasis on cultural construction rather than ma-terial determinacy can be seen as building on Elshtain's (1981) critique of socio-structural theories as 'narratives of closure'.

One of the major concerns of poststructural geographies, then, is the refutation of the notion of one single theory or 'grand narrative' capable of explaining spatiality throughout time and across space. In particular, poststructuralism denies the existence of one single truth or logical reason – logocentric constructs that have been so important in Western philos-ophy since the Enlightenment:

> the search for a single all-encompassing theory is therefore rejected in principle, as is the very possibility of objectivity. Western philos-ophy's quest for truth and certainty . . . is therefore abandoned and is seen as the product of a particular historical era that is becoming inappropriate in a postmodern society that is increasingly characterised by fragmentation, diversity and diffuseness in all spheres of life.
>
> (Bryson 1992: 225–6)

There is no doubt that contemporary social and cultural geographies have been influenced by the works of such key poststructural theorists as Foucault, Derrida, Baudrillard, Cixous, Kristeva and Irigary. Hartsock (1990: 169), however, expresses strong reservations about the usefulness of a Foucauldian analysis of power, stating that 'Foucault has made it very difficult to locate domination ... He has on the one hand claimed that individuals are constituted by power relations, but he has argued against their constitution by relations such as the domination of one group by another'. Hartsock therefore stresses the difficulty of embracing whole-sale the poststructuralist position because, she argues, poststructuralism's emphasis on social criticism as contextually, temporally and locally specific negates theories of power as systemic phenomena. With no systemic power there can be no overall system of domination and oppression, only specific contexts of subordination, resistance and transformation. Clearly, one of the questions for this book to examine is the extent to which systemic power exists in the creation of leisure and tourism landscapes and/or the extent to which localised, contextualised and pluralised power relations exert their influence upon the landscape. In attempting to answer such a question, the book seeks to situate contemporary leisure and tourism land-scapes within their historical contexts. Drawing on a range of sources and critiques, these 'histories' are problematised as social and cultural construc-tions of the past that offer neither total nor static explanations of the place of leisure and tourism as mediators of landscape.

The totalising explanations offered by the meta-narratives of structural theories are therefore seen as 'narratives of closure' because of their emphasis on power as repressive and dominating rather than productive and relational. Lash (1990: 4), developing Elshtain's critique, emphasises the importance of postmodernism and other cultural paradigms as 'regimes of signification'. These regimes, he contends, comprise the 'cultural economy' and the 'mode of signification'. According to Lash (1990: 5), the cultural economy is made up of the 'relations of production', the 'conditions of reception', the 'institutional framework that mediates between production and reception', and the 'particular way in which cultural objects circulate'. Drawing on established cultural studies analyses, Lash contends that the mode of signification depends

> on a particular relationship between signifier, signified and referrent
> ... Here, the signifier is a sound, image, word or statement; the signi-
> fied is a concept or meaning; and the referent is an object in the real
> world to which the signifier and signified connect.
>
> (Lash 1990: 5)

This book attempts to examine leisure and tourism landscapes as regimes of signification in which the production, representation and consumption of landscape are mediated by sites and processes of leisure and tourism. An analytical framework such as this, which recognises the productive and relational nature of landscape, is also more open to the concepts of resistance, subversion, transgression and reappropriation.

Social and cultural geographies of leisure and tourism landscapes

The book comprises ten chapters structured loosely into two parts. The first half of the book attempts to contextualise leisure and tourism landscapes by problematising the interrelations between landscape, leisure and tourism and by situating a discussion of these relationships within the wider discourses of social and cultural geographies. This first half of the book introduces a range of 'ways of seeing', representing and interpreting landscape. The second half of the book then seeks to examine ways in which dominant notions of landscape have been both maintained and disrupted by different processes of consumption, reproduction and representation.

Chapter 2, 'Locating Landscapes: Geographies of Leisure and Tourism', chronicles the development of geographical discourses that have informed our knowledge of the place of leisure and tourism within landscape studies and spatial theory more generally. This chapter 'sets the scene' by examining the place of landscape in creating both sites and sights of leisure and tourism. The chapter chronicles eight distinct geographical discourses that have informed our understanding of leisure and tourism landscapes. These discourses, in turn, inform many of the subsequent discussions within the book and are revisited in the final chapter where we attempt to relocate leisure and tourism landscapes. Chapter 3, 'Moving Landscapes: Leisure and Tourism in Time and Space', turns our attention to the contribution of historical geography in informing our understanding of the ways in which leisure and tourism landscapes have altered over time. This chapter highlights the significance of travel, the journey and the route to or through the leisure and tourism landscape. In drawing upon environmental studies, Chapter 4, 'Valuing the Countryside: Leisure, Tourism and the Rural Landscape', then examines the values that have been placed upon particular landscapes and assesses the ways in which such values have changed over time and across space. This chapter critiques the idealisation of the British countryside as a place of retreat and recreation for the benefit of urban visitors and tourists. Chapter 5, 'Representing Landscapes: Literary and Artistic Ways of Seeing', draws upon cultural geography and literary studies to inform our understanding of the ways in which dominant landscape values are represented through different media including poetry, prose, film and television. The chapter examines the way in which imaginative reconstruction, introduced in the previous chapter, has been used to construct a mythical Highland landscape in Scotland, a landscape that is central to the branding of Scotland as a tourist landscape.

In the second half of the book, Chapter 6, 'Heritage Landscapes: Merging Past and Present', traces the presence of the past within the landscape and reveals the contested notion of both heritage and history represented in contemporary leisure and tourism 'attractions'. Here, examples from the south-west of England illustrate the multi-vocal nature of Stonehenge, the evolving nature of Avebury and the mythical nature of heritage presented at Tintagel. Chapter 7, 'Gendered Landscapes: Constructing and Consuming Leisure and Tourism', continues the theme of the contested nature of space by examining the role that gender plays in the production, representation and consumption of leisure and tourism landscapes. Here, the focus on heritage landscapes is continued through an appraisal of a case study of the gendered representation of heritage in the Scottish town of Stirling. In Chapter 8, 'Retrophilia and the Urban Landscape: Reinterpreting the City', we then turn our attention to the representation of urban history for tourism consumption more generally. The focus of our case study in this chapter is London's City Fringe and the contested and reappropriated spaces within this urban arena. Chapter 9, 'Landscapes of Desire: Reappropriating the City', then draws our discussion of the contested nature of leisure and tourism landscapes to a close by focusing on the place of sexuality in constructing and consuming city spaces. Here, the chapter continues our discussion of the reappropriated nature of space in London by narrowing its focus from the city itself, to the street, and then to the environment of the hotel as a place of leisure and tourism.

Commensurate with a poststructural perspective, our journey through space and time continues beyond Chapter 9: there is no attempt at 'closure'. Rather, our final chapter, entitled 'Relocating Landscapes: Leisure, Tourism and Culture: in preference to 'Conclusion', attempts to synthesise the previous eight chapters by highlighting a number of common themes that emerge repeatedly throughout the course of our journey. These themes, informed by poststructural theory, are seen as significant in informing our knowledge of the interrelation between leisure, tourism and landscape. Moreover, these themes emphasise the importance of culture in shaping what have all too often been seen as the purely social or material relations of leisure and tourism. Each theme is testimony to the social–cultural nexus of the material and symbolic in the production, representation and consumption of leisure and tourism landscapes. As such, each theme has a place in the developing discourses of social and cultural geographies of leisure, tourism and culture.

2   Locating landscapes
Geographies of leisure and tourism

Introduction

The next chapter will highlight the interrelation between time and space in a *historical* analysis of the role of transport and communications in the construction, representation and consumption of leisure and tourism landscapes. This chapter, however, seeks to develop a *geographical* analysis of the place of landscape in creating both sites and sights of leisure and tourism.

Geographies of leisure and tourism have travelled almost full-circle over the last century, moving from spatial mappings of colonial territories and national boundaries to post-colonial discourses of power and the representation of place. Simultaneously, the scope and scale of geographical research has downsized from the macro-analysis of regional and national geographies to micro-analyses of everyday spaces, places and landscapes such as the street. Many of these more recent geographies have served to illuminate aspects of leisure and tourism research but have never been explicitly framed as leisure and tourism studies. This chapter appraises this broad range of developments by undertaking a chronological review of geographical research of leisure and tourism as landscapes of national identity through to streetscapes of leisure.

Alvesson and Billing (1997: 24) alert us to the dangers of uni-linear chronologies as descriptive frameworks when they state that 'all distinctions and ways of dividing up a complex, heterogeneous and rapidly expanding research area are problematic. They inscribe order and obscure disorder, ambiguity and variety'. There is, however, a purpose to such classifications as they seek to clarify the main focus and guiding principles of a body of knowledge or research while simultaneously attempting to situate the research area within a historical context. McDowell (1993a: 162), writing about feminist geography, has commented that 'Contemporary feminism is a diverse and pluralist project. Despite this warning, a chronological classificatory framework helps in understanding the changing emphases and context of feminist scholarship'.

This chapter outlines and evaluates a series of eight such geographical discourses that have informed research into leisure and tourism landscapes.

Each discourse is reflective of wider contemporaneous perspectives within social science and, as such, each is indicative of landscape theorising of the time. The review seeks to provide a historical overview of geographical analyses of leisure spaces, places and landscapes, progressing from accounts that stressed the physical, material and absolute nature of space to more recent analyses that emphasise the socio-cultural, symbolic and relative nature of space (Aitchison 1999a).

The relationship between geography and leisure research is well established (Barbier 1984; Carlson 1980; Mowl and Turner 1995). Indeed, the first issue of *Leisure Studies*, published in 1982, included an article by Coppock entitled 'Geographical contributions to the study of leisure' in which he chronicled the increasing involvement of geographers in the field of recreation and tourism since the 1930s. The article paid particular attention to methodological developments in the spatial analysis of recreation and tourism during the 1970s, the decade preceding publication. Since Coppock's article was written, however, the discipline of geography has undergone radical and transformative developments both resulting from and contributing to a number of new and competing philosophies within the social sciences. This chapter attempts to engage with these competing explanations of space and place in relation to leisure and tourism landscapes. In doing so, the chapter seeks to provide an introduction to the development of different discourses of landscape construction, representation and consumption. These discourses then serve to provide theoretical underpinnings to many of the discussions that follow in the subsequent chapters.

The first discourse outlined in this chronology is that of colonial geographies with their emphasis on a regional approach to mapping landscapes. The second discourse reflects systematic geographies that have attempted to model the pattern of landscape in relation to particular land uses. Third, the chronology outlines a series of geographical approaches to landscape evaluations and scenic amenity classifications related to countryside leisure and recreation. This third phase is then developed further in relation to tourism research within phase four of the chronology. Here, a series of tourism typologies is evaluated and their explanations of the development of tourism geographies are appraised. These first four phases of the chronology have previously been presented as largely apolitical, being informed more by economic concerns for landscape development. The fifth phase of the chronology, that of structuralist interpretations of landscape, introduces a more explicit political critique to geography's analysis of the development of leisure and tourism spaces and places. The chronology then identifies a series of recent post-positivist discourses of the production of space and place, and three specific schools of thought are discussed. The first of these is post-colonial geographies that offer a critique of the colonial discourse outlined at the start of this chronological discussion. Poststructuralist discourse is then developed in the next section when the impact of geography's cultural turn is evaluated in relation to the spatiality of leisure and tourism. This aspect of the chronology identifies synergies between a number of recent poststructural analyses of the spatiality of leisure and tourism landscapes. Three specific discourses of leisure and tourism geographies are discussed: leisure geographies of the street, tourism geographies of the monument and spectacle, and geographies of social and cultural exclusion. These poststructural discourses are then developed further in Chapters 7, 8 and 9. Chapter 7 examines the place of gender in constructing and contesting leisure and tourism spaces, places and landscapes. Chapter 8 explores the place of ethnicity in the production and reproduction of urban landscapes. Finally, Chapter 9 discusses the interrelation between sexuality and spatiality in city landscapes.

Colonial geographies: mapping regional territories

Whereas we can look to the Greeks and Romans for the origins of classical European geography, human geography is commonly perceived as being one of the more recent social sciences. The first university Chairs of geography were established in a number of European countries during the last two decades of the nineteenth century, and the last century saw enormous development and diversification within the discipline. Geography has relocated itself from the position of physical science at the beginning of the twentieth century, to social science in the middle part of the century, to that of a complex subject that mixes and merges concepts from the natural sciences, social sciences and humanities within the contemporary academy.

In the early part of the twentieth century geography was divided by two dichotomous relationships: physical and human geography, and regional and systematic geography (Mitchell and Murphy 1991). The early emphasis on physical geographical analysis was often developed in relation to specific regions of the world and coincided with the maturity of the nation-state, colonial expansion and imperialist interest in defining, mapping and characterising areas of the globe according to identified regional characteristics or differences. Indeed, Barnes and Gregory (1997: 16) point out that it can be argued that geography 'helped to provide a logistical basis for modern warfare, to foster a sense of national and imperial identity, to administer colonies, and to compile resource inventories for commercial exploitation'. In other words, geography did not develop as an impartial, detached, neutral set of perspectives through which 'objective knowledge' and 'truth' could be established. Instead, geography's *modus operandi* were inherently related to the legitimisation and reproduction of existing and emerging power structures.

These uncritical approaches, however, serve to construct place identities imbued with cultural meanings. Places may be represented differently across time and space, and even between different people at the same time and in the same place. Shurmer-Smith and Hannam (1994: 13) alert us to this view by emphasising that 'Place is a deceptively simple concept in geographical thought'. From a distance, places can be seen as foreign and exotic, while they may be familiar or even 'home' when viewed from close proximity. As the previous chapter indicated, notions of near and far have changed over time and the identities of places, and the people within these places, have also undergone transformation. Dualistic notions of near and far, home and away, local and distant, familiar and foreign then serve to construct that which is perceived to be known and unknown. The known places and people are those which we view as being similar to or the same as us, whereas those places and people with which we are un-familiar often take on the image of the Other. This concept of Othering, and its centrality to the creation of leisure and tourism landscapes, is discussed more fully in Chapter 7 and forms a continuous thread throughout the remaining chapters.

Systematic geographies: modelling land use and tourism

Following the First World War, the large-scale redrawing of the world map illustrated that human action was as much responsible for defining national boundaries as any physical landforms. The resultant shift towards human or cultural geography was simultaneously mirrored by a move from regional to systematic geography. This shift resulted in a number of regions being studied in relation to a particular phenomena or topic rather than studying a number of topics or phenomena in relation to a single region. Topics or phenomena that became established within the system-atic approach included agriculture, urbanisation and, more recently, tourism and recreation.

Within human geography, academics tend to define themselves as being interested in, or belonging to, a particular sub-discipline. Social, political, economic and historical geography all combine geography with another academic discipline to provide an interdisciplinary analysis of the relation-ships between people, places and the environment. Urban, environmental and cultural geographies offer a more multidisciplinary analysis of the social construction of space and place. More recently, feminist geography has pro-vided a systematic analysis of the spatial nature of patriarchal relationships within particular fields of human geography, whilst geography's cultural turn has emphasised the importance of gender, sexuality and disability in constructing spatiality. What unites all of these geographical analyses, how-ever, is their investigation of the relationships between people, places and the environment and the consequent interaction of society, space and nature.

Until the emergence of critical geographies in the late 1960s, there was a heavy reliance on mathematical and statistical modelling of land use within geographical research and theory (Coppock 1982: 4). Initially, however, these theoretical approaches were not concerned with leisure and tourism land use. Driven by the demands of maturing industrialised nations, geography's 'founding fathers' had little interest in developing leisure theory in the early decades of the twentieth century. For example, in 1909 Alfred Weber published his theory of industrial location suggesting that the optimum location for industry would be found where transport costs were at a minimum both for the inputs or raw materials and the finished product. Weber's preconditions included the availability of all information related to the location and production process, the existence of an isotropic plain – the geographer's substitute for the scientific laboratory – and the existence of single firms with only one plant.

Although Weber's model was widely used in determining the location of early twentieth-century seaside resorts, subsequent theories of 'time–space distanciation' (Giddens 1985) and 'time–space compression' (Harvey 1989a) were to demonstrate its anachronistic nature in an era of multinational tourism providers, relatively cheap and quick transportation, and global communications. The possibilities for rapid movement between distant spaces and the (re)presentation of the past within the present have formed the basis of much of the recent growth in leisure and tourism. Central to many of these developments has been the transformation of landscape for the purposes of leisure and tourism.

In 1940 August Losch attempted to define a model for the determination of the optimum geographical distribution of economic activity. Unlike Weber, Losch introduced the concept of competition but ignored production costs. Like Weber, he assumed an isotropic plain with firms of the same size behaving in the same way in order to maximise profit and deliver the same product to a market where customers had identical incomes and demands. This theory may have held some value in the 1940s when such mass leisure activities as cinema-going and cycling were experienced in similar ways and in similar proportions by a largely undifferentiated market. Today, however, our contemporary differentiated markets of leisure and tourism products, designed to meet the demands of self-styled lifestyle groups, have less to gain from such grand theories or meta-narratives.

Barnes and Gregory (1997: 1) refer to 'the hegemony of spatial science' when discussing the development of positivist geography in the post-1945 period. They claim that theoretical developments comprised the 'three C's': certainty, coherence, and cumulation in relation to the development of geographical knowledge. This early concentration on locational analysis and spatial science resulted in spatial determinism or a kind of geographical essentialism. Descriptive and predictive models were devised to explain the development of cultural geographical phenomena within a physical geographical context. Many of these models were to shape geographical thinking and research for decades to come and were strongly embedded within early recreation and tourism research within geography. While leisure studies was embracing sociological concepts from Marxist, structuralist and humanist theory in the 1960s and 1970s, tourism studies and recreation research appeared to hold on to their more positivist paradigm. Analyses of recreation and tourism continued to rely on quantitative methods to devise theories which mapped, modelled, coded and classified recreation and tourism development, provision, participation and impacts (Coppock 1982; Duffield and Owen 1970; Glyptis 1991; Patmore 1972, 1983). Glyptis (1991: 96) points to the difficulties of measuring demand for countryside recreation compared with that of leisure use at built facilities that have clearly defined entrance points. Consequently, many of the models developed to explain countryside recreation provision and participation are now seen as providing a rather narrowly defined analysis of recreation and tourism.

In 1960, William Alonso, building on Ratcliff's book *Urban Land Economics* (1943), provided a 'theory of the land market' which identified the concept of the bid rent curve. Bid rent theory defines a relationship between urban land use, location, accessibility, land value and economic utility so that urban spatial structure is determined by competition for scarce central sites which demand a higher rent than peripheral sites. Again, Alonso's theory was founded on a number of basic assumptions including the premise that accessibility and volume of business is likely to be greatest in the city centre and that transport costs increase as distance from the city centre is increased. In the 1970s, tourism geography attempted to adapt Alonso's model to explain the spatial differentiation of tourism development (Moissec 1977; Yokeno 1974). But the growth in such out-of-town leisure phenomena as cinemas, shopping centres and garden centres, combined with increasing inner city traffic congestion, raised a question mark over the applicability of such theories during the 1980s when urban decline was fuelled by a laissez-faire approach to planning. During the 1990s, however, something of an urban renaissance took place in a number of cities that have deliberately positioned themselves as places of cultural consumption (Hannigan 1998; Ward 1998). These developments are discussed in more detail in Chapter 8, where a case study of one such initiative in London outlines the changing nature of an area previously labelled as socially disadvantaged but now redesignated as a 'cultural quarter' in an attempt to boost both the social and economic regeneration of the area.

Landscape evaluations: mapping scenic amenity in leisure and tourism

The earliest geographical research of leisure and tourism landscapes was purely concerned with the classification of scenic quality and land use. These processes of land use mapping were instrumental in the subsequent designation of specific landscapes, space and places as sites of conservation and/or recreation. In 1934 Cornish published his article entitled, 'The scenic amenity of Great Britain' in which he aimed to provide a classificatory system for measuring scenic amenity of different landscapes. This mapping exercise relied upon databases of land use and vegetation type constructed by such people as Tansley (1939). In 1904 Tansley formed the British Vegetation Committee with the specific objective of mapping the vegetation of the British Isles (Green 1985: 45). The publication of Cornish's report also coincided with the establishment of the Standing Committee on National Parks by the Councils for the Preservation of Rural England, Scotland and Wales. The next fifteen years saw the publication of the Scott Report (1942), the Dower Report (1945) and the Hobhouse Report (1947), all of which contributed towards the establishment of a National Parks Commission following the National Parks and Access to the Countryside Act in 1949. In spite of these developments, however, Stamp (1948) did not include leisure, recreation or tourism as a principal land use when he attempted to classify and map land use across Britain. In the United States, however, a map of recreational land use had already been published, signifying the greater interest in the geography of outdoor recreation in North America (R. M. Brown 1935).

Some forty years later these mapping and modelling processes, together with other positivist measures of landscape evaluation, were rejected by the Countryside Commission for Scotland, which opted instead for the use of qualitative observational judgement to appraise scenic value (Countryside Commission for Scotland 1978). Crofts and Cooke (1974) and Blacksell and Gilg (1975) provide comparative critiques of the different positivist classifications of landscape value but only begin to recognise the concepts related to the social construction of space which were to receive such credibility twenty years later. In summary, while there is an extensive literature on areas such as national parks and their place as landscapes of leisure and tourism, the dominant discourse within such literature is one that seeks to map and model provision and participation within particular public policy frameworks. There has been limited interest in the subjective and socially constructed nature of landscape evaluation from recent social and cultural geography. In contrast, urban and commercial leisure landscapes have been scrutinised by contemporary social and cultural geographers, and these discussions are outlined later in this chapter and then developed further in subsequent chapters.

Tourism geographies: typologies of land use

By the late 1930s and early 1940s there was an increasing interest in the geography of seaside resorts, resort development and coastal landscapes as sites of leisure and tourism. Gilbert's 1939 paper entitled, 'The growth of inland and seaside health resorts in England' provided a systematic geographical approach to tourism development and planning. Economic geography provided a number of positivist models of industrial development and location which were adapted for use in tourism studies and which are still widely used today. Although many of the guiding principles of such models are well founded, the numerous preconditions attached to their application identifies their allegiance to the scientific method of research. Their durability was inevitably limited by a discipline that was later to question and then largely reject the use of positivism in human geography research.

More recent models, derived explicitly to explain tourism development, have often been founded on similar conceptual thinking to the positivist and predictive models of Weber, Losch and Alonso. The Ellis Curve predicts a progression from economic, social and cultural benefits to economic, social and environmental costs in the development of a tourist destination. R. W. Butler's 'Concept of a tourism area cycle of evolution' (1980) predicts the sequential movement from the exploration phase to involvement, development, consolidation and stagnation of a tourist resort. R. W. Butler's model, however, requires the existence of formulaic preconditions in the same way as Weber's, Losch's and Alonso's theories.

In addition to contributions from tourism economics and economic geography, research into the development of tourism landscapes and tourist destinations has also been informed by the psychology of tourism and tourism psychology. Plog, in his 1974 article titled 'Why destinations rise and fall in popularity', proposed a psychographic profile of tourist motivation in the form of a continuum that ranged from the 'allocentric' tourist at one end to the 'psychocentric' tourist at the other extreme. According to Plog, the allocentric tourist sought adventure and exploration whilst the psychocentric tourist preferred organised mass tourism or package holidays to popular destinations. Building on Plog's model, Butler and Waldbrook (1991) offered a parallel continuum of tourism impacts that ranged from the low environmental, social and economic impact of allocentric travellers to the high impact of psychocentric tourists. Previously Cohen (1979) had also examined the interrelation between tourists and destinations pointing out that the 'drifter' would have less impact than the 'organised mass tourist'. Such assertions, however, are now beginning to be questioned by tourism research contending that tourists who are constrained within the 'tourist bubble' provided by the mass tour operator actually inflict less impact upon local populations and landscapes than the independent tourist, traveller or backpacker. In other words, there is a 'need to distinguish between sustainable tourism and the development of tourism on the principles of sustainable development' (Butler 1999).

Structuralist interpretations of leisure and tourism landscapes

By the 1960s, the focus of geographical research had begun to shift and a more critical approach was becoming evident. Gregory, Martin and Smith (1994: 3) assert that this transitionary period signalled a move 'away from a preoccupation with the statistical laws of spatial distribution to the historical and material processes of uneven urban and regional development inherent in the "laws of motion" and "crisis tendencies" of capitalism'.

Structuralism seeks explanations for observed phenomena in general structures that underpin all phenomena but are not identifiable within them. With the structure as construct theory, observed phenomena are seen as representations of deep structures genetically imprinted on human consciousness. However, structuralism can also be seen as structure as process where phenomena are seen as representations of underlying social structures. Here, it is the transformation of structure at societal rather than neural level that is significant. Such theory has formed the basis of Marx's work, of the Marxian humanism of the Frankfurt School, the critical theory of Habermas (1989) and Gidden's theory of structuration (1984).

Traditional structuralist theory therefore emphasises the importance of societal structure and constraint, whilst phenomenological theory emphasises the importance of human agency. In contrast to this seeming polarisation, social theorists such as Berger (1963) and Giddens (1984) point to the mutual dependence of both structure and agency, such that a dialectic is formed between individuals and society with each influencing the other in a continuous process. Thus, the individual both shapes and is shaped by society. As such, cultural meanings are often built into the structures of institutions, the symbols and cultural artefacts of society, and the landscapes of the built environment and rural surroundings. Many of the leisure and tourism landscapes discussed in subsequent chapters illustrate these dialectic relationships between structures and cultures or between society and the individual.

During the 1960s and 1970s Marxist analyses had re-emerged in other disciplines through the works of Thompson (1968) in social history, Althusser (1971) in social theory and R. Williams (1961, 1965, 1977) in literary criticism and cultural studies. Geography followed this trend with Harvey producing some of the most notable work in the form of *Explanation in Geography* (1969) and *Social Justice and the City* (1973). Harvey attempted to explain the spatial structure of the urban landscape by applying a Marxist analysis to production and consumption patterns. Production and capital accumulation, according to structuralist perspectives, are intrinsically linked to the labour market process and the design and function of urban areas. As capitalism has developed, and cities have grown in size and complexity, there has been a shift from manufacturing to service industries as the major urban economic activity. This, in turn, has transformed the urban landscape from being one of production to one of near-simultaneous production and consumption. Later sections of this chapter, together with Chapters 7, 8 and 9, seek to examine the spatial organisation of leisure consumption within the social and economic fabric of the city. Furthermore, it is in the urban locale at the turn of the millennium that the combined forces of leisure, culture and commerce appear to have most impact in shaping personal and social identities and these processes and performances will be evaluated further in Chapters 7, 8 and 9.

The Marxist analyses outlined above were not, however, extended to leisure and tourism geography as this field remained dominated by the positivist paradigm. Moreover, although the late 1970s and early 1980s were to be identified as a period of common reorientation as a result of the impact of Marxist and structuralist theory within many of geography's sub-disciplines, any continuity was short-lived. A series of new approaches, labelled variously as post-Marxist, poststructuralist, critical-realist, structurationist, feminist and postmodernist, were soon to pose challenges to the newly established structuralist orthodoxy.

As outlined in the Introduction, Elshtain (1981) has described structuralist or socio-structural theories as 'narratives of closure' in reference to their totalising explanations of social phenomena as products of material relations. These theories, emphasising the social and material nature of power, often explain power relations in terms of their repression and domination of social groups. In contrast, socio-cultural theories stress that power is 'productive, relational and everywhere' (Cooper 1994). Analyses that focus on the cultural and symbolic nature of power are more receptive to the concepts of resistance, transgression and disruption, particularly in relation to the re-negotiation of power relations by marginal groups. The next three sections of this chapter examine some of these forms of disruption in relation to leisure and tourism landscapes. These discussions are then continued in relation to gender, ethnicity and sexuality in Chapters 7, 8 and 9 respectively.

Post-colonial geographies of leisure and tourism

Post-colonial geographies can be viewed as part of a wider series of post-positivist geographies including postmodernism and poststructuralism (Barnes and Gregory 1997: 2). Although the focus of each post-positivist geography is different, they embrace a commonality in their rejection of the grand narratives and totalising visions of truth and objectivity that were portrayed by the previous positivist geography. It is from this era of post-positivist geography that the new cultural geography emerged and became merged with sociological and cultural studies analyses to investigate the multiplicity of behaviours, meanings, consumption trends and identities constructed in and through leisure and tourism. Indeed, Sibley (1995: 137) asserts that this emphasis on difference and devolution of power has posed a problem for 'a subject whose history is so much bound up with colonialism'.

Post-colonial theory enables us to reflect critically upon constructions of leisure and tourism destinations as foreign and exotic and the people within these landscapes as Other. This perspective also renders possible a critique of the legacy of colonialism that is manifest in many of the structures and practices of contemporary leisure and tourism (Memmi 1967; Said 1978). The previous sections of this chapter have highlighted the dominance by Western academics of geographical theory. Post-colonialism, with its focus on the global rather than the Western or parochial, has challenged these authorial voices and the representativeness of white, Western, and usually male thinking. Post-colonial theory has been instrumental in questioning the homogenising categories of hosts and guests, so fundamental in much of the tourism literature. Memmi (1967: 85), writing about 'the colonised' alerted us to 'the mark of the plural' (all Others look the same). Said (1978, 1993) and Spivak (1985, 1987, 1993) then examined the role of language and discourse in representing and sustaining Other identities and the power relations inherent in the production and consumption of Other identities. Shurmer-Smith and Hannam have pointed to the link between geography's imperialist legacy and contemporary tourism:

> Military geography's masculine gaze on other cultures was often enshrined through that other instrument of power, photography ... People in foreign countries were often assembled for their portraits to be taken for geography texts as if they were exhibits rather than human beings.
>
> (Shurmer-Smith and Hannam 1994: 18)

Perhaps today we have substituted the tourist brochure for the geography text. Edwards (1996: 197), in her analysis of 'Postcards: greetings from another world', examines 'the role of modern tourist postcards of "ethnographic" subjects in perpetuating notions of the exotic'. Images of 'the aborigine', 'the native', 'the Indian', 'the village chief' or 'the village elder' form a substantial part of Edwards' collection of 1,500 modern ethnographic postcards housed at the Pitt Rivers Museum in Oxford. Edwards argues that the production and consumption of traditional culture as tourist commodity now forms an integral part of the tourism industry. The increasing ease of travel to what, until only recently, were defined as long-haul destinations provides both stimulation and demand for these images of Others. A passage from Unijet's 1999–2000 Wintersun brochure summarises these socially constructed tourist desires only too well:

Sri Lanka is simply stunning. Shaped like a jewel adorning the Indian sub-continent, it's no surprise to find the island is so rich in rubies and sapphires. But Sri Lanka's real gems are its people, its food, its breathtaking beaches, spectacular highlands and its fascinating culture. In such continuous bright sunshine, the island's colour contrasts are startling – white sandy beaches are fringed with deep green palm trees, giving way in turn to the even deeper textures of the tea plantations and paddy fields. Temples and shrines are found everywhere, often overgrown and overrun with local wildlife which adds to their mysterious, spiritual atmosphere. In a land where tourists become travellers, where a holiday becomes an experience, it's hard to believe that you're only eleven hours flying time from the UK.

(Unijet 1999: n.p.)

While MacCannell (1976) explained the desire for such images as the tourist's search for the authentic, more recent tourism commentators are more cynical. Building on the work of Boorstin (1964), Turner and Ash (1975) reflect on the possibility that tourists seek out, not authenticity, but the pseudo-event. Visiting the monument and the spectacle, or gazing on the staged authenticity of Turkish belly dancing or Indian Kathakali dancing specially adapted for tourists, is seen as part and parcel of the contemporary tourist experience. Drawing on the first-hand experience of local novelists and journalists, it is possible to offer a post-colonial critique which, if offered in our own words, would only be the words of academics, viewed through the eyes of tourists, engaged in the process of Othering. For example, Roy (1997), in *The God of Small Things*, refers to the colonialist influence of contemporary tourism and its impact on local and regional culture such as the performance of all-night Kathakali dancing in Kerala where the original form of the dancing has been corrupted for the benefit of Western tourists:

In Ayemena they danced to jettison their humiliation in the Heart of Darkness. Their truncated swimming pool performances. Their turning to tourism to stave off starvation.

On their way back from the Heart of Darkness, they stopped at the temple to ask pardon of their gods. To apologise for corrupting their stories. For encashing their identities. Misappropriating their lives . . .

. . . In despair he turns to tourism. He enters the market. He hawks the only thing he owns. The stories that his body can tell.

He becomes a Regional Flavour.

In the Heart of Darkness they mock him with their lolling nakedness and their imported attention spans. He checks his rage and dances for them. He collects his fee. He gets drunk. Or smokes a joint. Good Kerala grass. It makes him laugh. Then he stops at the Ayemenem

Temple, he and the others with him, and they dance to ask pardon of the gods.

(Roy 1997: 229–30)

What the spectre of staged authenticity does, of course, is to present the possibility of multiple realities and contested meanings. Here we turn to contemporary social and cultural geographies to provide insight into the processes by which leisure and tourism spaces, places and landscapes are constructed, contested, disrupted and transformed as a result of competing and changing patterns of consumption and identity formation.

Geography's cultural turn: the spatiality of leisure and tourism

Soja (1985) introduced the concept of spatiality to emphasise the socially produced and interpreted nature of space. Contemporary social and cultural geographies demonstrate that space, place and landscape – including landscapes of leisure and tourism – act as sites and sights of social and cultural inclusion/exclusion and are not fixed but are in a constant state of transition (Urry 1995). These spatial transformations result from continuous, dialectical struggles of power and resistance among and between the diversity of landscape providers, users and mediators. As such, recent social and cultural geographies place emphasis upon both agency and structure but are simultaneously critical of the work of previous human geographers who 'celebrated' place and failed to problematise space and place in relation to socio-cultural constructs such as gender, sexuality, race, class and (dis)ability (Gleeson 1999).

Giddens's (1984) theory of structuration, and the interplay or dialectic between structure and agency, can certainly be identified as influential to these social and cultural geographies. In this recent geographical work, however, structuration theory is widened to embrace the diversity and plurality of agencies, to include the role of agencies previously defined as marginal or peripheral, and to recognise the significance of micro social and cultural phenomena and subcultures in shaping spatial relations. Current social and cultural geographies therefore provide a meeting ground for the anti-foundationalist movements of postmodernism, poststructuralism and post-colonialism. Their analyses attempt to embrace the complexity of spatiality rather than engaging in the spatial determinism or essentialism of the early twentieth century or the grand narratives of structuralism from a decade or so ago. As Gregory, Martin and Smith (1994: 5) emphasise, 'geography's task is now seen to involve dialogue with other disciplines rather than instruction of them, to produce complication as much as simplification, and to issue admissions of doubt as often as declarations of certainty'.

This task is in stark contrast to that emphasised by Barnes and Gregory

(1997) and outlined at the start of this chapter where certainty of empirical observations, coherence of patterns, forms and processes, and cumulation of knowledge and discovery were seen as central tenets of a systematic science. This emphasis is not shared by contemporary social and cultural geographies where difference and diversity are frequently revealed in the nuances of marginal social and cultural phenomena and spatial transformations.

Theories emphasising the social *and* cultural construction of landscape have played a central role in establishing the discourse of contemporary social and cultural geographies. Over the last decade, a number of prominent geographers have illustrated that, rather than being a physical or objective reality, 'landscape is a cultural image, a pictorial way of representing, structuring and symbolising surroundings' (Daniels and Cosgrove 1988: 1). Landscape is seen to represent a social and cultural geography of the imagination (Gregory 1994) and this point is illustrated aptly by Rose who states, 'Whether written or painted, grown or built, a landscape's meanings draw on the cultural codes of the society for which it was made' (1993: 89). These codes are embedded in social relations and frequently reflected in cultural symbols. Theorisation of the relationship between society and culture by social and cultural geographers has so far drawn on the humanist Marxist tradition of Gramsci (1985), Williams (1977), Thompson (1968) and Berger (1963), all of whom saw the material and symbolic dimensions of the production and reproduction of society as inextricably intertwined.

Recent social and cultural geographies caution against prioritising the cultural over the social or the 'unharnessing of the symbolic and the sociological', as identified by Bondi (1992a: 166). Shifts from structural to cultural analyses within many of the social sciences have led some commentators to view poststructuralism, with its emphasis on the symbolic, as having unharnessed itself from any theoretical perspective which provides a sufficiently rigorous analysis of social, material and systemic power. Indeed Gregory, Martin and Smith (1994) contend that:

> in prioritising the mapping of fragments or differences, a postmodern human geography runs the risk of ignoring the more systematic features and relations of social, political and economic structure: the geographical imagination may well be broadened in one direction, but is simultaneously eclipsed in another.
>
> (Gregory, Martin and Smith 1994: 10)

The challenge for social and cultural geographies of leisure and tourism landscapes, then, is to provide a broad analysis of the cultural 'fragments and differences' in the interrelation between these social and cultural phenomena whilst simultaneously attending to broader structural relations of power and place. A number of geographers have recently attempted to achieve just such synergy between structural and cultural analyses of space, place and landscape in relation to sites and processes of leisure and tourism. Although these geographies have rarely been framed explicitly as leisure geographies or geographies of leisure, a brief summary of a sample of recent research and writing serves to demonstrate the centrality of leisure and tourism to contemporary social and cultural geographies. The following discussion is structured around two very different types of land-scapes that have accommodated leisure and tourism: the landscape of the street, and the landscape of the monument and spectacle. Both of these types of leisure and tourism spaces or places can then be contextualised as landscapes of social and cultural exclusion and are discussed further in relation to gender and sexuality in Chapters 7 and 9 respectively.

Leisure geographies of the street

Geographies of the street have shifted from viewing the street as a place of residence or business to a place of leisure, consumption, identity forma-tion, fashion, spectacle and performance. In the introduction to *Images of the Street*, Fyfe (1998: 1) describes streets as 'the terrain of social encounters and political protest, sites of domination and resistance, places of pleasure and anxiety'. Rendell (1998: 75) demonstrates how, even in the early nineteenth century, the street became a site for the 'spatial repre-sentation of gendered identities'. Wearing and Wearing (1996) and Wearing (1998: 131–35) point to the street as a leisure space for the contemporary chorister and the *flaneur*, who have used this public space to display and perform their social and cultural identities. Their work draws on de Certeau's notion of 'space as practised place' where, 'the street defined by urban planning is the place which becomes transformed into space by the people who use it' (de Certeau 1984: 117). Poststructural analyses, embraced by contemporary social and cultural geographies, offer an insight into the contested, negotiated and renegotiated nature of the street as a site of leisure and tourism. The perspective of poststructuralism presents a number of ways of theorising these urban landscapes and two concepts, both central to the work of Foucault are particularly insightful here: the gaze and surveillance.

Earlier, this chapter referred to the construction of the Other within leisure and tourism geographies. As a means of viewing the Other, tourists employ *the gaze* and references to *the tourist gaze* entered the discourse of tourism studies following the publication of Urry's book of that title in 1990. Previously the concept of 'the gaze' was more commonly discussed within poststructuralist theory by writers such as Foucault (1976: 89). Central to the concept of the 'gaze' is the theory that knowledge is both socially constructed and socially constructing. The power of the gaze is unequally distributed and the object of the gaze is constructed according to the locus of power and control. Theories accentuating the social–cultural nexus emphasise that, through the act of the gaze, people, 'differently engaged and differently empowered, appropriate and contest their landscapes' in different ways (Bender 1993: 17):

> Landscapes are thus polysemic, and not so much artefact as in process of construction and reconstruction ... The landscape is never inert, people engage with it, re-work it, appropriate and contest it. It is part of the way in which identities are created and disputed, whether as individual, group, or nation-state.
>
> (Bender 1993b: 3)

Foucault also introduced us to the concept of social and cultural surveillance. The introduction of closed circuit television (CCTV) has increased the surveillance capacity of modern cities (Rule 1973). Surveillance of the street has, according to Fyfe and Bannister (1998: 263), contributed to the privatisation, purification and homogenisation of public space. This perspective sees that the proliferation of activities which are deemed counter to the interests of business, civic pride or city residents are arrested, both metaphorically and literally, as a result of the gaze of the camera. The prohibition of certain types of leisure-related behaviour, such as consuming alcohol in the street, is given greater enforcement under the watchful eye of CCTV. Miles (1997: 197) refers to the 'spread of surveillance culture' within urban streets and Foucault went as far as to suggest that this regulation of public space was also a regulator of bodies and body space. We are perhaps more familiar with the presence of surveillance cameras in our indoor streets or shopping malls and M. Davis (1990: 240) refers to the contemporary shopping centre as 'the mall-as-panoptican-prison' which presents an interesting interplay between the concepts of the gaze and surveillance theorised in *Discipline and Punish* (Foucault 1977). There is a counter discourse, however, that views CCTV as a protector of diversity. The presence of the cameras can be used for protective as well as punitive policing. Proponents of the use of CCTV point to its central role in policing racist attacks in our city centres and, more recently, to catching the homophobic and racist bomber who targeted shoppers in Brixton's main street and the gay pub *The Admiral Nelson* in Soho in 1999. Indeed, the use of CCTV has almost become a marketing feature in place promotion. Cirencester, for example, informs visitors arriving by road that they are entering 'Cirencester: now a safer town with CCTV'.

Tourism geographies of the monument and spectacle

The monument and spectacle have received increasing scrutiny from social and cultural geographers, in addition to extensive critiques from tourism scholars. Rojek (1993), in his discussion of tourist attractions as 'escape areas', identified four types of tourism 'escape areas' and each embodies either the monument or the spectacle: 'heritage attractions', 'black spots', 'literary landscapes' and 'theme parks'. The use of the monument and the spectacle as tourist attraction has traditionally served to maintain or replicate cultural symbols of colonial domination, or to represent Others through the deployment of the tourist gaze. More recently, however, there has been increasing recognition of the place of landscape art as a type of monument from and to the people. Within these landscapes, like the streetscapes discussed above, leisure and tourism have played significant roles in the emergent culture of near simultaneous production and consumption. This section seeks to illustrate discussions of the power of representation and the representation of power within tourism landscapes by making reference to a range of monuments, spectacles and public art forms that have become part of the landscape of contemporary tourism.

Barthes (1982), in *Empire of Signs*, alerted us to the representation of power and the power of representation displayed by monuments, statues, signs and symbols in the modern city. In this way, any critique of material artefacts and symbols, such as statues and monuments, has to be undertaken within a wider social and cultural geography which acknowledges the social relations of the society in which the artefact was produced and the cultural context within which it is displayed. As Chapter 8, 'Retrophilia and the Urban Landscape: Reinterpreting the City', will go on to discuss, monuments rely upon a commonality of interpretation by those who gaze upon them. As Miles (1997: 58) contends: 'Monuments are produced within a dominant framework of values, as elements in the construction of a national history ... they suppose at least a partial consensus of values, without which their narrative could not be recognised'. This recognition of a common narrative relies upon what Nietzsche (1980) identified as monumental and antiquarian historiography. A monumental historiography consists of a history written, or in this case built, in such a way that past events, people and places are presented or represented as icons to be celebrated and embraced as providing role models for the future. Both the construction of the Wallace Monument in Scotland, and the representation of Wallace within the monument, discussed in more detail in Chapter 7, can be seen as an example of monumental historiography. Wallace is represented as a symbol and icon of Scottish nationalism that has endured for over 700 years, providing not just a glimpse of the past but a key to future national freedom. Further examples of the monumentalist approach can be seen in Nazi Germany, where monuments were erected in celebration of some mythical past but also as a means of stirring nationalist fervour for the future. In contrast, an antiquarian historiography is more conservative in that it seeks to preserve the past without any immediate aim of using the past to influence the future. The conservationist and preservationist policies of organisations such as the

National Trust, discussed in Chapters 4 and 6, provide examples of anti-quarian historiographies. Heidegger (1980), however, has stated that all historiographies, rather than just monumentalism, are concerned with the future. The difference between historiographies, he argues, is merely in their attitude towards the future. Thus, monumentalism uses the past to embrace the future whereas antiquarianism is much more tentative in its approach towards the future.

This chapter commenced with a discussion of colonial geographies and it is from that same era that the majority of British monuments were erected to commemorate colonial expansion and military power. It is not just a historical accident that statues of Queen Victoria became ubiquitous throughout Britain's major cities towards the end of last century. The four decades following Victoria's death then witnessed two world wars and the quashing of numerous anti-colonialist uprisings, many of which are cast in stone and bronze at focal points in urban public places. Shurmer-Smith and Hannam (1994: 203) proclaim these monuments as 'deliberate physical manifestations of ideology, a more or less massive inscription of triumphal or laudatory statements upon the landscape'. Many of these celebrated 'triumphs' reflect acts of violence, domination and oppression and both D. Mitchell (1992) and Miles (1997) have suggested that the centrality of violence to the monument is equally central to the maintenance of hegemony through the construction of collective national memories. Chapter 7, 'Gendered Landscapes: Constructing and Consuming Leisure and Tourism', explores this concept in greater detail by focusing on the example of Scottish nationalist monuments in Stirling.

If social and cultural geographies teach us anything, however, it is that power can be viewed as contested and fluid. The spatial representation and manifestation of power can change over time, be disrupted across space, and even transgressed within the same space or in the betweenness of space. Alongside the national monuments invoking collective memory from the top down, we now have a range of monuments and public art forms that celebrate localised collective memory and local resistance to forms of national collective memory. Miles (1997: 76) refers to the democratisation of the monument, drawing on the examples of Raymond Mason's *Forward* in Centenary Square, Birmingham and Kevin Atherton's *Platforms Piece* at Brixton railway station. Both sculptures are intended to reflect local people rather than national icons. Such art forms, or democratised monuments, may even represent animals rather than people. From *Greyfriars Bobby* in Edinburgh to *Snooks the Dog* in Aldeburgh, the erection of monuments to local, and (usually) heroic, dogs provides a link between local history or mythology and the British collective national memory related to 'man's best friend'.

Geographies of social and cultural exclusion

Many of these democratised monuments have sought to redress the exclusion of local people from leisure and tourism landscapes and can also be used to reflect local resistances and uprisings to outside influences, or to reflect local pride and identity. Robertson (1998), for example, has recounted the recent construction of a series of monuments on the Island of Lewis in Scotland's Western Isles. Here, islanders have worked together, under the guidance of a local architect, to build monuments out of local stone that commemorate local uprisings and resistances against domination from previous landlords. Similarly, Anthony Gormley's 65-foot steel *Angel of the North*, currently the UK's tallest sculpture, was erected overlooking Gateshead as a guardian angel and celebration of a previous era of heavy industry including local coal mining, steel making and shipbuilding. A further Tyneside sculpture, dubbed the *Easel of the North* in reference to its steel tripod frame, designed by Mark di Suvero and titled *Tyne Anew* will soon become the UK's largest piece of public art and a further evocation of Tyneside's industrial heritage. Di Suvero (quoted in Hetherington 1999: 11) has stated that he 'wanted to add to this feeling of the past impacting on the present. The roots of shipbuilding and coal-mining are important to the area and must never be lost sight of'.

The publication of *Arts and Sport: A Report to the Social Exclusion Unit* (Department of Culture, Media and Sport 1999) offered further recognition 'that leisure has an important role in meeting many of the wider social objectives that contribute to social inclusion, community development and neighbourhood renewal' (Local Government Association 1999: 1). *Arts and Sport* (1999) was intended to report on 'best practice in using arts, sport and leisure to engage people in poor neighbourhoods, particularly those who may feel most excluded, such as disaffected young people and people from ethnic minorities' (Department of Culture, Media and Sport 1999: 5). The Policy Action Team for Sport and the Arts (PAT 10) was one of eighteen such teams established following the publication of the Social Exclusion Unit's report *Bringing Britain Together: A National Strategy for Neighbourhood Renewal* (Social Exclusion Unit 1998).

This focus on urban regeneration and neighbourhood renewal has much in common with established areas of interest within urban and social geographies (Gregory and Urry 1985; T. Hall 1998; Harvey 1973, 1985a, 1985b, 1989a; Jackson and Smith 1984; Johnston 1984; Ley 1983; Pahl 1970). Leisure studies, too, have a tradition of considering the role of sport, the arts, leisure and recreation in community development (Aitchison 2000a; Bramham 1989; Coalter, Long and Duffield 1988; Haywood 1994; Henry 1997; McDonald and Tungatt 1992; Voase 1997). More recently, tourism studies have also considered the interrelation between tourism development and urban renaissance in a number of British towns and cities (Bianchini and Parkinson 1993; Richards 1995, 1996; Urry 1990; Wynne 1992).

Policy Action Team 10's remit of identifying ways in which leisure can contribute to neighbourhood renewal therefore provides both an academic opportunity and a political rationale for identifying synergy between geographical research and leisure research. For example, the DCMS report stated that:

> We also want those involved in the arts and sport and those involved in regeneration to recognise that they can contribute greatly to each other's aims by working together. Arts and sports bodies should acknowledge that social inclusion is part of their business. Equally, area regeneration schemes should explicitly incorporate arts and sport in neighbourhood renewal. Arts and sport are not just an 'add-on' to regeneration work. They are fundamental to community involvement and ownership of any regeneration initiative when they offer means of positive engagement in tune with local interests.
>
> (Department of Culture, Media and Sport 1999: 6)

Inevitably, however, policy rather than theory has dominated the political discourse of leisure and social exclusion and the report serves to identify persistent discontinuity between theory and policy. This is particularly so in relation to the distinction between recent theoretical developments embodied in social and cultural geographies and more traditional leisure-related policies. It has been argued elsewhere that this policy-related discourse has frequently verged on rhetoric and has offered a rather simplistic view of the role of leisure, sport and the arts in relation to urban renewal (Aitchison 2000b). Indeed, the Local Government Association has recently voiced concern that 'much of the evidence surrounding this debate seems to be piece-meal and mainly anecdotal' (Local Government Association 1999: 1). In spite of the recency of both the DCMS and the Social Exclusion Unit, this discourse or rhetoric itself is not new. Following the English urban riots of 1981, a number of initiatives such as Action Sport were launched in inner city areas as a means of using sport and leisure to combat social unrest, particularly among young unemployed people. These policies were later widened in the form of the Sport for All projects and targeted at specific sectors of the population deemed to be underrepresented in sports participation: women, people over 50 and disabled people. Almost two decades later a similar rationale for developing sport and the arts as a means of tackling social exclusion in urban areas is voiced in the DCMS report:

> Arts and sport are inclusive and can contribute to neighbourhood renewal. They can build confidence and encourage strong community groups. However, these benefits are frequently overlooked both by some providers of arts and sport facilities and programmes and by those involved in area regeneration programmes.
>
> (Department of Culture, Media and Sport 1999: 5)

But this rhetoric of equality and social inclusion, based upon a dualism of inclusion/exclusion, has served to construct notions of a single dominant culture or community to which those excluded are presumed to aspire. These constructions emphasise hierarchies of difference and can serve to militate against notions of cultural diversity and heterogeneity. The inclusion/exclusion dualism thus appears to imply that to move from being excluded to being included requires excluded individuals and groups to adopt and conform to a predetermined set of roles, behaviours and values already in place. This hegemonic discourse has been accommodated within both leisure policy and leisure studies, but clearly poses problems for social and cultural geographies underpinned by poststructural theory and an emphasis on difference and diversity (Aitchison 2000c).

Overview

Both Urry (1995) and Shields (1991) have stressed the importance of spatiality in constructing leisure patterns and relations whilst Rojek (1995: 1) has emphasised the importance of cultural analysis to our understanding of leisure in late modernity by stating that 'one cannot separate leisure from the rest of life and claim that it has unique laws ... the object of leisure is subsumed by the subject of culture'.

The apparent lack of articulation between geography and leisure studies seems particularly surprising at a time when we are witnessing the increasing breakdown of disciplinary boundaries within the academy and when social and cultural geographies are engaging with a range of analyses that focus on leisure and tourism sites as key social and cultural signifiers. For example, recent research has contributed to our understanding of leisure and tourism spaces ranging in scale from the proximity of body space to the intangibility of cyber space (Crang, Crang and May 1999; Duncan 1996). *En route* we have encountered urban space, performance space, transitory space and liminal space. Leisure sites that have formed the basis of such research have included the gym, the pub, the club, the holiday, the theme park, the monument, the festival and the parade, among others (Bell and Binnie 1998; Nast and Pile 1998; Shurmar-Smith and Hannam 1994; Skelton and Valentine 1997).

Sibley (1995) in *Geographies of Exclusion* drew our attention to his analysis of the spatial domination and power exercised over both private and public spaces and places. Although neither leisure nor tourism feature in the index to his book, Sibley's critique makes extensive reference to the ways in which both leisure and tourism have served to construct spatial relations from the shopping mall to the public park. Similarly, Shields' (1991) analysis of *Places on the Margin* addresses the ways in which previously marginalised groups have appropriated peripheral spaces and places as their own, frequently through the activities or sites of leisure and tourism. Fyfe's (1998) collection of *Images of the Street* again assesses ways in which exclusion is both effected and resisted in public places, and the myriad ways in which informal leisure activities serve to construct and deconstruct notions of place. In addition, Sibley, Shields and Fyfe all make reference to gender and sexuality as exclusionary markers and mechanisms in space, place and landscape. Chapter 7 will develop these explanations and critiques of gendered leisure and tourism landscapes while Chapter 9 will address issues of sexuality in the construction and representation of urban landscapes. In addition, Chapter 8, in its examination of 'Retrophilia and the Urban Landscape: Reinterpreting the City', discusses the place of ethnicity in the production and consumption of urban space. These *Geographies of Exclusion* (Sibley 1995) are then revisited in the final chapter: 'Relocating Landscapes: Leisure, Tourism and Culture'.

3   Moving landscapes
Leisure and tourism in time and space

Introduction

Recent academic literature has given considerable emphasis to the processes through which everyday working places are transformed into alluring landscapes of leisure and tourism. Less consideration, however, has been given to the appreciation of scenes observed *en route* to the destination. The experience of passing through an ever-changing landscape enables the traveller to catch glimpses of places to which he or she has no lasting commitment. The journey can be regarded not simply as a means to an end but as a source of pleasure derived from a sense of freedom through travel.

To someone undertaking travel just three or four centuries ago, however, the idea of taking pleasure in the journey itself would have seemed strange indeed. Today, it is not easy to appreciate the difficulties and dangers encountered by people making journeys which now seem comparatively short. The old spelling of *travaile* – used by some authors until the early 1700s – makes a telling association between movement, hard work and physical discomfort (Wallace 1993). The vast majority of people spent their lives confined within the perimeter of a day's walk from home. Those who tramped the roads were generally treated with suspicion. The poor wayfarer was a marginal figure, a potential thief, subversive or social parasite (Heal 1990).

For more prosperous citizens, a nationwide system of transport and hospitality developed in response to market demand. Nevertheless, their diaries and travelogues suggest that the passing scenery held little appeal for the majority of cultivated travellers until well into the eighteenth century (Parks 1964). Until that time, journeys through such wild and rugged terrain as the Lake District, Snowdonia or the Highlands of Scotland, provoked few positive comments from those who travelled on horseback, by private carriage or stagecoach (Thomas 1983). But, from the mid- to late 1700s many among the leisured classes acquired a taste for travel to view awe-inspiring natural scenery, the ruins of ancient buildings, as well as new manufacturing plants that could be seen as monuments to the nation's ascendant industrial and military power (Moir 1964). By the early nineteenth century, such itineraries were well established, if not well worn. Movements in taste have thus led to emergent patterns of travel, which in their turn have been expanded and systematised into a tourist industry (Ousby 1990: 5).

The ideology of those with wealth and power tended to reinforce the notion that such circulation or touring remained their prerogative, whereas the immobility of the poor confirmed their subordination. To the rising middle classes of the nineteenth century, journeying to places which came to symbolise the nation and its heritage conferred inclusion and possession. The itinerary could express patriotism and identification with the interests of the state. This privileged sense of belonging could also be implied through the acquisition and perusal of landscape prints: panoramic views which provided an analogue for travel (Helsinger 1994). By the mid-nineteenth century, however, the distinction between the mobile and immobile population had been disturbed by a transport revolution which brought a sharp fall in the cost of travel, as well as some spectacular improvements in speed and convenience (Bagwell 1974). Remote regions, once visited with difficulty and discomfort by the few, became accessible to the many.

As Towner (1996: 11) notes, tourism studies research has sometimes adopted a less than critical stance on the significance of transport as a determining factor in the evolution and pattern of tourism. It is more realistic, perhaps, to conceptualise its role as enabling in the context of the changing social, economic and political system of which it is a part. Nevertheless, as Urry (1995) argues, movement is central to the very idea of modernity. Mobility is 'responsible for altering how people experience the modern world, changing both their forms of subjectivity and sociability and their aesthetic appreciation of nature, landscapes, townscapes and other societies' (1995: 144). This chapter examines the development of travel to consume the scenic attractions of the British Isles, and explores relationships between freedom of movement, social mobility and power. It outlines the historical development of transport that facilitated the growth of domestic tourism, and discusses the various ways in which the places observed along the way have been represented, communicated and reproduced.

The journey, travel and discovery

The early modern period saw an increasing number of people and vehicles on the move, especially in England. During this period of transition from feudalism to capitalism, Webb and Webb (1913: 63–4) note the association between economic growth and a rise 'in the number of well-mounted horsemen, with their saddle bags behind them'. Wealthy travellers could ride post, changing their mounts at staging points along the route. But even by this costly method the pace was often slow and robbery remained a significant hazard. As yet, there were no signposts, and in most counties the standard of highway construction and repair was poor because the work was carried out with understandable reluctance by unpaid surveyors and parish labourers. Furthermore, the main thoroughfares had deteriorated since medieval times with the increase in traffic, churned by pack horses and animals being driven to market and rutted by a rising number of wheeled vehicles. In the 1500s, heavy road wagons were introduced to carry freight for hire and reward. These were large hooded vehicles pulled by eight horses, covering 15 miles on a good summer's day. From the late sixteenth century there were many more private carriages, and in the early 1600s stage coach services began. In 1622 the waterman poet John Taylor lamented 'suddenly, this is a rattling, rowling, rumbling age' in which 'the world runs on wheels' (quoted in Webb and Webb 1913: 69).

The expansion of trade also created a rising demand for accommodation at inns and taverns. By 1500, these provided a convenient system of lodgings for travellers along the main overland routes that radiated out from London, the seat of government, international port and mercantile metropolis. Over the next century, provision expanded to cover most of England at nodal points of the road network from the major cities out to the more remote and smaller market towns. In the 1540s, following the Dissolution of the Monasteries, the former town residences of abbots and priors – most of which had rooms furnished to lodge pilgrims – were sold off by the Crown, and many became large inns. Those at Glastonbury, Lewes, Malmesbury, Peterborough and Cirencester became the Dolphin, Walnut Tree, Castle, Bell and Popingay Inns respectively (Inwood 1998: 154). In neighbouring countries, including Scotland, commercial hospitality remained far less developed for many years to come, and the standard of that available in England was often praised by foreign visitors (Heal 1990).

Most of those who patronised such accommodation were merchants and people travelling on public service, civic or personal business. Long journeys for reasons other than necessity were the exception: riding across Britain on horseback in the sixteenth and seventeenth centuries was still *travaile*. In medieval times, the ritual of pious travel to a holy shrine had been undertaken as an act of penance or devotion or, as Wallace (1993) intimates, a search for knowledge and understanding. Chaucer's *The Canterbury Tales* suggests, however, that the spiritual journey had not been without its worldly distractions. In England, the Reformation replaced papal doctrine with a radical new Protestant theology which *inter alia* brought an abrupt end to religious pilgrimages. The Tudor monarchs and their successors nevertheless encouraged a secular version of travel which came to be known in Europe as the Grand Tour. The state needed an educated governing class of noblemen and gentlemen, endowed with knowledge of international politics, as well as an appreciation of the languages and customs of the European mainland. This 'moving academy' consequently became an important source of prestige and social capital.

England was an emerging power and cartographers were mapping out a new geography of commerce, warfare and diplomacy. The medieval *Mappa Mundi* had shown a flat Earth represented by a circle centred on Jerusalem, with the distorted shape of the British Isles at the very edge of Christendom. Now, the Renaissance world view held sway. European explorers had reached China, colonisation of the New World had begun and the British Isles were no longer on the world's periphery. The nation-state became Great Britain after union with the monarchy of Scotland in 1603, and its economic and military strength continued to increase. As wealthy British citizens embarked on tours of mainland Europe, so foreign visitors came to view the sights of London, Oxford, Cambridge and the royal palaces. As yet, however, travel on home ground was not particularly fashionable. Some graduates of the two- to three-year Grand Tour might spend a month or so exploring their own country, but in general this lacked the status, intellectual stimulation and excitement of foreign travel.

The increasing number of well-mounted horsemen and the hazards they encountered on Britain's highways, nevertheless helped to create a market for local maps, route guides and topographic surveys of the nation. These became important features of travel culture, and interpretation of their language can provide useful insights. As Harley (1988) argues, they have never been value-free representations of place. It is therefore helpful to consider maps and surveys in the context of their production and use by educated travellers who could interpret their meaning. In keeping with contemporary social and cultural geographical analysis, recognition should be given to that which may lie below the literal sense. Content is selective and the signs and styles of representation are 'ways of conceiving, articulating and structuring the human world which is biased towards, promoted by, and exerts influence upon particular sets of social relations' (Harley 1988: 278). They may therefore highlight aspects of the world which map-makers consider worthy of inclusion, as well as emphasising cartographic silences.

By the late sixteenth century, professional surveyors commissioned by the landed gentry had perfected the accurate mapping of estates. Some used the techniques of their trade to produce county maps which could be reproduced in quantity. Copper plate engraving had replaced woodblocks, permitting much finer detail and a larger format for maps. In the 1570s, Christopher Saxon used this process to publish an impressive series of thirty-four English county maps. These showed the location of towns and villages, as well as rivers, woods and hills, but omitted the road network. In the 1590s, John Norden's maps included the main thoroughfares which connected settlements, with useful devices such as inset town plans, and scales in the margin. Nearly a century later, John Ogilby made a further advance in the form of large folios of strip-maps enabling riders to plan their route. From 1719, the demand for more portable road-books for use on the journey itself was satisfied by Thomas Gardner's *Pocket Guide to the English Traveller*. Ousby comments on the significance of these developments:

> The road itself is emphasised at the expense of natural features like hills and rivers, which, unless they constitute an impediment to the journey, are relegated to the same category as country houses or churches marked along the margins of the route, as sights or features of interest falling within the traveller's view. In a sense, these strip-maps embody the triumph of tourism: what is known, or worth knowing about a place, is what can be seen on the route to somewhere else.
>
> (Ousby 1990: 14–18)

The relative attractions of the Grand Tour and physical hardship of domestic travel notwithstanding, the seventeenth century marked a pioneering phase in the discovery of Britain. A small but influential minority of dilettante and topographic writers were beginning to champion the patriotic virtues of touring their own country, fired by an impulse for knowledge and desire to observe first hand. Moir (1964: 13) discusses the enthusiasm of topographers 'to write descriptions of their native land, the fruits of their own travel and observations, glorying in her past because of pride in her present greatness'. The style of their regional histories and geographies owed something to descriptions of foreign countries, particularly those for use on the Grand Tour, with manuscript or print substituting for a tutor. Their purpose was to educate and inspire, providing the reader with a *Speculum Britanniae*, a looking glass of Britain to mirror faithfully the country and all its achievements.

During the seventeenth century, some travel writers paid tribute to Britain's history as evidenced by ancient remains, such as stone circles, hill forts and ruined castles. They also honoured the nation's contemporary achievements – new mines, furnaces, shipyards and the draining of the fens – now landscapes of industrial tourism. The published diaries of Celia Fiennes' journeys, written 1685– *c*. 1712, provide some keen observations of what the author deemed significant. Born into a military family that had supported parliament in the Civil War, she was a self-proclaimed enthusiast for healthy exercise, education and pleasure through travel on home soil. Fiennes advised her readers:

> that if all persons, both Ladies, much more Gentlemen, would spend some of their tyme in Journeys to visit their native land, and be curious to inform themselves and make observations of the pleasant

prospects, good buildings, different produces and manufactures of each place, with the variety of sports and recreations they are adapt to, would be a sovereign remedy to cure or preserve from these epidemick diseases of vapours, should I add Laziness?

(quoted in Morris 1982: 32)

She concluded with 'a hearty wish and recommendations to all, but especially my own Sex, the studdy of those things which tends to improve the mind and makes our Lives pleasant and comfortable as well as proffitable in all Stages and stations in our Lives' (quoted in Morris 1982: 32–3).

Fiennes rode side-saddle on extended itineraries through England and Wales, and her energetic narrative conveys an admiration for enterprise and improvement. Liverpool, for example, was 'a very rich trading town where the houses are of brick and stone built high and even, that a streete quite through lookes very handsome' (quoted in Morris 1982: 161). In marked contrast to the working centres of commerce and industry, Fiennes informed her readers of the watering places whose main business was the leisure of their clientele. Fresh water had become a valued commodity in expanding urban centres, especially the metropolis, and medicinal springs and wells within a day's walk were popular places of recreation, with pleasure gardens and entertainment such as Thomas Sadler's musick house in Islington. These tended to be at the lower end of a social hierarchy of resorts and were described by her contemporary Daniel Defoe (1970) in *A Tour Through the Whole Island of Great Britain*, first published in 1712:

> as the nobility and gentry go to Tunbridge, the merchants and rich citizens to Epsome; so the common people go chiefly to Dullwich and Stretham; and rather also, because it lies so near London, that they can walk to it in the morning and return at night; which abundance do ... which makes the better sort also decline the place; the crowd on those days being both unruly and unmannerly.
>
> (Daniel Defoe 1970: 166)

Specialised settlements which owed their existence to the health-giving qualities of mineral water springs had existed since Roman times, but it was the rising prosperity of the leisured classes and their desire for diversions as well as health cures that stimulated the development of spas, and later seaside resorts. As yet, few who travelled across country to the fashionable watering places, or to enjoy the hospitality of the great houses on country estates, had much to say about the scenery along the way. Some took delight in observing land improved by agriculture, but there was a marked antipathy towards the wild and rugged landscapes which most modern tourists find so appealing. On visiting the Lake District, Celia Fiennes was disturbed to find herself 'walled in on both sides by those inaccessible high rocky barren hills which hangs over ones head in some places and appear very terrible' (quoted in Morris 1982: 168). Thomas (1983) has explained this in terms of the anthropocentric world view: the sterile uselessness of wilderness contrasted with the utility and beauty of land brought into cultivation. Furthermore, there was a lingering association between mountains and divine punishment. This was expressed as late as 1681 in Bishop Gilbert Burnet's *Sacred Theory of the Earth*, an endorsement of the doctrine that the Earth's surface had been smooth and uniform, unperturbed by sea or mountain, until God's wrath was vented on humankind through the Great Flood (Nicholson 1959).

Prospects of pleasure, landscapes of feeling

Between 1700 and 1800, however, important new relationships were to develop between the idea of landscape and an emerging culture of travel. *Landskip* had first entered the English language in the late sixteenth century from the Dutch as a painterly word to describe the pictorial representation of a piece of countryside as a subject in its own right or as the background scenery behind a portrait (Barrell 1972). Once regarded as an inferior art, over the first half of the eighteenth century landscape art became highly respected and the dominant style was Italian. European culture was deeply influenced by Renaissance Italy where the artistic innovation of linear perspective froze reality at a specific moment 'removed from the flux of time and change, and rendered the property of the observer' (Cosgrove 1998: 22). A fine collection of landscape paintings evinced affluence and good taste, and the works of Claude Lorraine, Nicholas Poussin and Salvator Rosa were especially prized. These recalled the pastoral idylls of the Augustan poets: nostalgic evocations of the Roman *campagna* of a Golden Age, inhabited by gods, goddesses, nymphs and other figures of mythology. The new sensibility favoured the gentle melancholy of ancient ruins, crumbling and overgrown to harmonise with Arcadian themes (Andrews 1989; Lowenthal 1985).

In the last quarter of the eighteenth century, the aesthetic of the Picturesque became a fashionable way of contemplating the actuality of scenery, consciously derived from painters' compositions. Hitherto, art could imitate a landscape. Now, a landscape could imitate art. The major influences included Claude, Ruisdael and other members of the Dutch school, as well as the quintessentially English landscapes of Thomas Gainsborough, whose rustic scenes such as *The Woodcutter's Return* (1773) and *The Watering Place* (1777) portrayed a deep countryside, untouched by the agricultural improvements and enclosures which were by now changing the face of much of the Midlands and Eastern England. There was a preference for the asymmetric, for rough and rocky terrain, winding lanes and watercourses, shady trees and natural vegetation, broken gates, well-worn stiles and dilapidated and derelict buildings (Bermingham

1986). Viewed from a distance, figures in the landscape could also be Picturesque: low life characters poignantly adorned with tattered and ragged clothes going about their daily lives, or in repose watching over livestock, their posture deferential.

The Picturesque gaze required detached observation to visualise the harmonious whole, an abstraction which was made easier by the Claude glass. In origin an optical device used by painters to assist composition, typically it comprised a portable convex mirror about four inches across, mounted on black foil and in a convenient travelling case. The viewer sat with his or her back to the true landscape, observing its modified reflection in the mirror. The miniaturised, reflected image gave detailed definition only in the foreground and distorted the background so that its features shrank away into the distance (Bell 1993). This had the effect of superimposing on the reality of British landscapes effects corresponding to a Claudian type: neo-classical in form, composition, colour and tone. However, landscape taste was shifting, as Greek and Roman symbolism was repudiated or naturalised (Andrews 1989). Italianate pastorals bowed to 'humbler English rural scenery [which] represented a landscape both familiar and accessible. It thus could be widely *consumed,* and with all the more enthusiasm in that the landscape it celebrated was beginning to vanish' (Bermingham 1986: 85).

Engravings satisfied a wider market: 'looking over prints' became a genteel afternoon pastime for the drawing room and sketching landscapes an outdoor activity for the well mannered. An appreciation of scenic beauty in reality, as well as in art, was an important social accomplishment, rather like the ability to sing well or to compose a polite letter (Barrell 1972: 5). It required knowledge of artistic and literary allusions, as well as an educated eye to frame the view from the correct vantage point or prospect, dividing up foreground, middle ground, far distance and sidescreens. Soon, through printed text illustrated by engravings, some authoritative tutors became widely known, and the Reverend William Gilpin, a retired school teacher and parson, was one who attracted many followers. As a clergyman Gilpin did not deny that nature was great in design, but in detail 'she' often fell short of the ideal portrayed in art. The work of humankind, such as ruined abbeys, could also disappoint. Sketchers were therefore licensed to use discretion, making minor improvements as they toured the countryside in search of fine prospects. The line was drawn at introducing new features such as imaginary castles, but the composition could be enhanced by planting a few additional trees, removing a fence, modifying the course of a river or road, and so on (Moir 1964: 124).

Gilpin's *Observations on the River Wye, and Several Parts of South Wales* (1782) expounded the principles of landscape appreciation in a region that demonstrated the necessary qualities, subject to imperfections in nature. This volume was followed by a further seven, illustrated by mezzotints of the author's pen and wash sketches. Other places so commended ranged from the coasts of Hampshire, Sussex and Kent to the Highlands of Scotland. Thomas West's (1780) *Guide to the Lakes*, with its inclusion of poet Thomas Gray's contemplations in the second edition, was also influential in promoting the region's attractions. Disruption of the Grand Tour by war and revolution on the continent proved a further boon, and the Lakes became a playground for the prosperous. The pioneering stage of discovery thus gave way to a second phase in which a tourist industry developed to satisfy demands that varied from the practical to the whimsical. These included accommodation, guided walks, boat trips, the firing of cannons to echo around the mountains, dances and musical *soirées* and a museum of curiosities. As yet, most visitors were content to admire the scenery from the lower slopes. Steps were cut into the hillside with markers to indicate the stations or viewpoints recommended by West and other writers. Villas also appeared, their landscaped gardens and parkland representing a notable modification of the lakeside scene (Towner 1996: 151).

By the time that mainstream bourgeois culture was so readily absorbing the ethos of Picturesque taste, the sensibility of the intellectual avant-garde was, however, moving in a new direction. In conscious reaction to the Enlightenment, with its emphasis on reason and progress, Romanticism was inspired by the philosophy of Rousseau and Goethe and the mysticism of William Blake. In art, the formal rules of neo-classicism were eschewed in favour of spontaneity and freedom of the imagination. Kroeber (1975: 29–31) discusses the contribution of John Constable, whose painting *The Cornfield* (1826) depicts a locality and moment charged with meaning for the artist. In the foreground, a boy drinks from a pool below a group of trees, one of which is old and decaying, and the eye is led to clouds which move across the sky. The middle ground is interrupted by a sheepdog driving a flock along the lane below towards a fallen gate, a plough and two farm labourers working in a cornfield. The viewer's journey leads to low hills and a church in the distance: perhaps a glimpse of the life beyond. For Constable, it was the experience and memory of the scene which mattered, and a landscape that moved the emotions to such an extent could sometimes be too beautiful to imitate on canvas (Honour 1979: 68).

The discourse of Romanticism renounced the anaesthetising comforts of urban life, repudiating the materialism of an industrialising society. Wallace (1993) examines the origins of walking long distances to refresh the mind, body and soul. Such journeys allowed participants direct encounters with nature and a feeling for the essence of place. In the late eighteenth and early nineteenth century, the virtues of excursive walking were advocated and practised by William Wordsworth and his companions who formed the vanguard of artistic and literary taste. Roaming the countryside on foot had previously been associated with impoverished wayfarers who had no other means of conveyance. Now, the idea of walking purposefully but with no practical aim became a form of recreation enjoyed by those with access to alternative means of transport had they wished to use it. Through the 1800s, as travel became less physically demanding, faster and cheaper with a wider choice of modes, the practice of the Romantics became an established leisure pursuit.

Annihilating time and space

Britain was in the early stages of a transport revolution. Industrialisation and the market economy for farm produce required improvements to the trunk thoroughfares across the country and by 1800 parliament had authorised over 1,600 turnpike trusts to raise capital, upgrade roads, install bars or gates and charge regulated tolls (Bagwell 1974: 38–9). Many remote areas were still poorly served by local roads. For example, the vogue for Picturesque touring, which peaked in the 1790s, predated significant highway improvements within most of the wild and rugged areas visited. But there was now a compelling economic rationale to link smaller settlements to market towns and hence to regional centres and the metropolis. Barrell (1972: 84) suggests that these developments should be understood as part of a wider movement 'to *explain* the countryside, open it out, and to make each particular place more available to those outside it'. There is therefore a concurrence between this process and Gilpin's way of seeing the landscape, with its desire to fix and appropriate the scene:

> The concern to be always moving through a place, to see it never primarily as a place-in-itself, but always as mediated by its connection to one place in the east, and another to the west, produces a sense of space which is defined always by this linear movement, so that to stop at a place is still to be in a state of potential motion.
>
> (Barrell 1972: 89)

This way of seeing landscape was also a significant feature of the work of J. M. W. Turner. Daniels (1993: 112) comments that his 'pictures of places are not merely local in their meaning. They situate places in their regional, national and international contexts.' These representations could be interpreted as a meeting of the everyday and the epic. Thoroughfares and intersections of routeways, including rivers, canals and sea lanes, turnpikes, drove-roads, and later railways, are prominent elements in some of Turner's most popular compositions. Development of the nation's industry, transport infrastructure and the increasing bustle of traffic could be viewed with pride by the ascendant middle classes, and engravings of works by Turner and other less-celebrated painters were reproduced singly or in bound volumes. Such engravings promoted a kind of vicarious tourism in that they offered circulation and access to the greatness of the nation. In the Napoleonic wars, patriotic sentiment expressed through landscape art reached new heights and Turner's contribution included a set of prints illustrating the bulwark shore entitled *Picturesque Views of the Southern Coast of England* and published in 1811.

In some of his later work such as *Picturesque Views in England and Wales,* published in two volumes in 1832 and 1838, Turner's figures were, however, inclined to break with the Picturesque convention of depicting the rural poor occupied in useful toil, engaging pity and subservient to the composition and the viewer. In contrast, his foregrounds were sometimes filled by working class people, themselves engaged in leisure at fayres, courting, drinking, dancing or strewing the ground with bottles and other litter. Critics, including John Ruskin, chastised him for the 'vulgarity' of such trangressive figures which 'defaced' his paintings, reminding the onlooker of 'the coexistence in the same space of multiple cultures, undermining the concept of a single aesthetic nation constituted through landscape viewing' (Helsinger 1994: 118).

The development of rail transport had far-reaching implications for leisure activities across a broad social spectrum. From the 1830s, cheap travel to the coast, countryside and festivals was facilitated by charter trains and excursions. Horse racing, prize fighting and other activities now attracted even larger crowds and became mass spectator sports. And, with rail access, some seaside resorts grew at a remarkable pace, catering for daytrippers as well as wealthier, longer-term guests accommodated in hotels, together with retired and other permanent residents. Granville (1841), introducing his comprehensive guide to England's spas and seaside resorts, hailed the railway as a means of conveyance which seemed to level all topographical distinctions, annihilating time and distance. In 1851, visitors by the million were brought to London by excursion train to celebrate the nation's industrial prowess at the Great Exhibition. Limited purchasing power and free time remained, however, significant constraints for many who experienced overcrowded urban living conditions, including women and children whose leisure was generally confined to the street and local neighbourhood (Walvin 1978).

The building of the railways represented a feat of heroic engineering. Management of the workforce required iron discipline to synchronise activities and operate timetabled services in safety at unprecedented speeds (Shaw 1993). At first, individual companies instituted their own time convention. Later, this was standardised, so that time itself was redefined by the new imperative: the precision of Greenwich Mean Time replacing local reckoning (Schivelbusch 1986: 43). In *Dombey and Son,* for example, Dickens (1848) refers to the pervasiveness of the new railway culture:

> There were railway hotels, coffee-houses, lodging-houses, boarding-houses; railway plans, maps, views, wrappers, bottles, sandwich-boxes

and timetables; railway omnibuses, railway streets and buildings, railway hangers-on and parasites, and flatterers out of all calculation. There was even railway time observed in clocks, as if the sun itself had given in.

(quoted in Richards and MacKenzie 1986: 2)

Urry (1995) emphasises the central importance of the development of rail transport during the second half of the nineteenth century in redefining relationships between nature, time and space. The very building of the railways flattened and subdued nature. The traveller, propelled through space, was separated as never before from the moving landscape which 'came to be viewed as a swiftly passing series of framed panoramas' (Urry 1995: 119). Schivelbusch (1986) argues that rail travel put an end to the intense experience of traversed space characteristic of the pre-industrial age, and highlights contemporary use of the word 'projectile' to describe the rapid movement of the train. To the passengers, hurled forward at speeds of up to 75 miles per hour, the track, cuttings and tunnels appeared as the barrel through which the missile was fired. The view from the carriages was neither to the front nor back, but to the side, an evanescent landscape to which they were no longer connected. As velocity dissolved the foreground, train travellers felt as if they no longer belonged to the same space as the perceived objects which passed by so quickly. Many found difficulty in relating to this novel perception of landscape. Relating to fellow travellers was also a problem to which the habit of reading on long train journeys provided only a partial solution.

Itineraries geared to this faster pace of travel demanded a new type of guide to places which could be visited by train. Ousby (1990) refers to Martineau's (1855) 'brisk railway-age' companion to the Lake District, and to others who interpreted the literary landscape such as Scott in Scotland, Kingsley in Devon, and so on, 'making it by far the easiest way for tourists to convince themselves that they were feeling appropriate or edifying emotions in the presence of a particular scene without necessarily spending much time actually looking at it' (Ousby 1990: 187). Guidebook culture was also significant in the later development of the South West as a destination of superior social tone, through collaboration between coastal resorts and the Great Western Railway (Shaw and Williams 1991; Thornton 1993). Echoing our later discussions of the representation of a mythical Scotland, the imagery conveyed a Celtic Otherness, a 'reinvention' of Cornwall which was largely a creation of outsiders. Thus, the sub-region 'began to be viewed through an exaggerated prism of romantic bourgeois sensibilities ... It was becoming picturesque, its towns and villages quaint, its people moody, mystical, superstitious and childlike by turns' (Deacon and Payton 1993: 71).

New relationships were developing between town and country. No longer a part of everyday life for the majority, the idea of rural Britain as a place to which the town-weary might escape in their leisure time was further accentuated. Suburban living in greener surroundings was becoming the norm for middle-class Victorians and the countryside was an easy journey by train and omnibus. On the one hand, the rural landscape became the subject of earnest scientific enquiry, its geomorphology and natural history studied and classified. On the other, it became a realm of narrative and symbolism. Romanticised through the art of British landscape painters, the countryside was now a stage on which emotional drama could be acted out, safely distanced by dressing its actors in rural costumes, historic or contemporary. Themes explored by the Pre-Raphaelite Brotherhood, and highly popular with those who came to view their work, included financial ruin, emigration, sexual desire and transgression, illness and death (Bermingham 1986).

Ford Madox Brown surprised the Victorian art world by depicting suburban scenes, raised from the quotidian through subtlety of colour, microscopic attention to detail, and intensity of mood. *An English Autumn Afternoon, Hampstead – Scenery in 1853* provides a spectacle of modern life in its depiction of two young lovers with a view over back gardens, orchards and rooftops. Bermingham (1986: 175) contrasts it with the Hampstead views that Constable had painted in the 1820s. Constable's scenes had orchestrated the tones from dark to light, organising the space into receding planes, and selecting features for special emphasis. Brown's Hampstead, by comparison, is neither tonally nor chromatically composed: everything is shown in sharp focus with nothing picked out for special attention. Such scenery was typical of the semi-rural landscape on the fringe of large cities, but hitherto an unusual choice for a painting. John Ruskin asked 'What made you take such a very ugly subject?' to which Brown retorted 'Because it lay out of a back window' (Treuherz 1993: 88–9).

The 1850s brought a new fashion in British genre painting to depict contemporary middle-class life where travel and leisure activities were represented in such places as omnibuses or trains, aboard ships, in parks, on beaches or at the races. William Powell Frith's panoramic crowd scenes included *The Railway Station* (1862), *Life at the Seaside* (1854) and *The Derby Day* (1858). Exhibited, these became major attractions in their own right with *The Derby Day* attracting over 21,000 people in seven weeks and all three attaining still greater renown as engravings. Indeed, they became so well known that they were imitated on stage in popular melodramas during which the actors froze in tableau to resemble the pictures (Treuherz 1993: 105–7). The mass reproduction of these and other scenic icons, such as works of Sir Edwin Landseer, was made possible by the new process of printing from steel plates. Furthermore, the plates themselves could be duplicated, legitimately or otherwise, through stereotyping. Tens of thousands of prints could thus be produced for the mass market of middle-class art consumers (Pringle 1988).

Coinciding with the early development of rail transport, photography soon established its position as a creator of images which defined the desirable scene. Up until the turn of the century, this was the domain of professional art photographers or highly committed amateurs. These were individuals who tended to adopt the bohemian lifestyle associated with Romantic artists, maintaining their superiority over mere tourists. As J. Taylor (1990) explains, 'taking' a landscape with wetplate technology required technical know-how and expensive equipment. It also involved a field expedition or 'campaign' with a tent and a strong sense of mission:

> tourists were said to be satisfied if they could see a beauty spot 'tolerably well' from the road. It was the business of the photographer to 'scramble' for the 'superb view' – which might then be turned into a picture-postcard. The idle tourists would buy this, not because it represented their view but because it was the ideal view.
>
> (Taylor 1990: 181)

Since the Edenic landscape could be saved in photographs it was possible to avoid the countryside altogether yet still identify with the rural ideal as a vital feature of national life. In this respect, publications such as *The Homes and Haunts of Shakespeare* (1894), and *One Hundred Gems of English Scenery, A Collection of Permanent Photographs* (1902) tapped the market for surrogate travel which had been established through engravings. The viewer gained admission to the nation's heritage: 'those whose tastes or means do not allow such visits may be in a position to surround themselves, at merely nominal costs, with faithful presentments of quiet scenic beauty, such as can never be excelled by any country in the world' (quoted in Taylor 1990: 182). By the early 1900s, photography was itself becoming accessible to amateur snap-shooters of modest means. Kodak had pioneered the production of cheap hand-held cameras, along with the industrialisation of film processing and printing. Sontag (1973) comments on the close association between this development and the rise of mass tourism:

> For the first time in history, large numbers of people regularly travel out of their habitual environments for short periods of time. It seems positively unnatural to travel for pleasure without taking a camera along. Photographs will offer indisputable evidence that the trip was made, that the programme was carried out, that fun was had.
>
> (Sontag 1973: 9)

It is generally accepted that the new medium played a role in the decline of figurative painting, yet the camera's unilinear view 'allowed the conventional landscape way of seeing, the visual appropriation of nature as a commodity, to be sustained and disseminated among a population far broader than that which landscape painting had addressed' (Cosgrove 1998: 257–8). Through its 'realistic' representation of places near and far, captured at a specific moment, photography also played its part in the transformation of people's consciousness of space and time.

Walsh (1992) reviews other significant inventions of the nineteenth century. In 1800, information could move at the speed of the fastest horse and by mid-century at the speed of a train. The electric telegraph, introduced by railways to send messages between signal boxes, led to the development of public telegraph systems which permitted near-instantaneous communication. Invention of the telephone in the 1870s facilitated long-distance transmission of the human voice and by the 1890s telegraph cables were being laid below the ground or on the sea bed. By 1900 it was possible to assimilate information concerning events and people across and between continents with little or no time lag: 'The development of mass communications permitted a sense of synchronicity as many people consumed information from people and places, sometimes almost instantaneously' (Walsh 1992: 66).

Landscape, leisure and mobility

By the turn of the century, mass-produced bicycles had also become available to artisans and the lower middle classes. With improvements to roads leading out of urban areas, a cycling craze caught on. This new form of recreation could be enjoyed by the lone rider but the camaraderie of the local cycling club was also a key feature of the fashion, especially among young men (Towner 1996: 238). The bicycle became an emblem of social progress that inspired a network of socialist Clarion Clubs (Rubinstein 1977). It also became an expression of emancipation for the New Woman, a trend that gave offence to some. Writing in the *Lady's Realm* (1897), Eliza Linton expressed abhorrence at this 'new development in feminine freedom' and the 'intoxication which comes with unfettered liberty' (quoted in Rubinstein 1977: 62). The democratic appeal of cycling was, however, to be endorsed by no less a figure than Lord Balfour, Prime Minister: 'There has not been a more civilising invention in the memory of the present generation ... open to all classes, enjoyed by both sexes and all ages' (quoted in Pimlott 1947: 166). Cycling also provided freedom from the constraints of public transport routes and timetables and, in this respect, foreshadowed the appeal of the private car (Pimlott 1947).

As yet, motor cars provided a novel means of touring available only to adventurous and prosperous individuals such as Hermann Lea, inventor, photographer, cyclist, water-diviner, bee-keeper and pioneering automobilist. From 1898, Lea developed a close friendship with Thomas Hardy, with whom he cycled and motored many miles across the West Country, exploring and photographing the landscapes that featured in the author's work (G. Cox 1986). Through this collaboration, Lea's (1913) guide was published as a companion volume to Macmillan's definitive editions of Hardy: a *vade-mecum* to interpret the literary landscape of rural life through touring maps and annotated photographs. Gruffudd (1994) comments on the aesthetic dialogue between tradition and modernity, facilitated by an unprecedented growth in personal mobility over the first half of this century. He notes Poet Laureate Alfred Austin's quest for 'Old England' (1902) and how he 'came across a countryside which had the abiding charm of ancientness and a population which pleased him, as it had eighteenth and nineteenth century travellers, with its servility and its moral sense' (Gruffudd 1994: 249).

Roads were, nevertheless, soon to be upgraded and traffic signs were erected in response to the increasing volume and speed of motor traffic. Charabanc tours and a network of bus services reached out into rural areas. Enthusiasm for the private car, and the still greater freedom and speed of movement that it offered, was reflected in a plethora of guides and motoring maps for those wishing to explore Britain's heritage and countryside. Writer J. B. Priestley (1935) attests to the pleasures of exploring the countryside slowly on foot or bicycle, 'sipping' rather than 'gulping' the landscape. However, in his introduction to a Batsford guidebook *The Beauty of Britain,* he also describes the allure of faster travel:

> Speed is not one of my gods; rather one of my devils; but one must give the devil his due. I believe that swift motion across a countryside does not necessarily take away all apprehension of its charm. It depends on the nature of the country ... There is a certain kind of dullish, rolling country, not very attractive to the walker or slow traveller, that becomes alive if you go quickly across it, for it is turned into a kind of sculptured landscape. As your car rushes along the rolling roads, it is as if you were passing over a relief map.
>
> (Priestley 1935: 2–3)

Although photography replaced some figurative art, such as place-illustrations in the text of guidebooks, landscape was to remain a compelling theme for British painters. In the earlier decades of the century, their diverse influences included Impressionism, Expressionism, Cubism, Vorticism and Futurism. The qualities of space, structure and motion were explored and representation of landscape veered between topographic realism and abstract use of colour, tone and texture. Their subjects were equally wide-ranging. Pleasurable scenes of the seaside by Laura Knight and Philip Wilson Steer depict children playing in peace and security. These provide a stark contrast with the trauma of foreign battlefields portrayed by Paul Nash, Sydney Carline and other 'war' artists. Walter Sickert, Charles Ginner and Joan Eardley are best known for their urban or industrial scenes whereas artists such as Ben Nicholson with his views of the Romney coast and of Cornwall and F. C. B. Cadell with his representations of Iona both depicted wild and elemental places devoid of human habitation.

Some demonstrated considerable versatility in their use of various media. Landscapes were represented in oil, pastel and watercolour, wood engravings, textiles, silk screen prints and lithographs, not all of which were to be shown as exhibition work. Art found commercial sponsors, including railway companies whose travel posters were prominently displayed to all who passed through stations and other public places, with these images reaching much larger audiences than visitors to galleries. Well-known artists were also commissioned to design dust jackets and other artwork for book publishers and notable contributors to the inter-war heritage and countryside genre discussed above included Eric Ravilious and Brian Cook. Landscape was also an inspiration to sculptors. The curvatures of Henry Moore's monumental reclining female figures echo the contours of the spaces they occupy, and Barbara Hepworth created a series of abstract wooden sculptures, 'some of them extremely sensual, intended to evoke the clean curves of the Cornish landscape' (Graham-Dixon 1996: 217).

The moving images of film made it possible not only to project visions of spectacular landscapes onto a cinema screen but to convey the sensation of moving through panoramic scenes that could be set in the past, present or future. With reference to the imaginary future landscapes of H. G. Wells' *Things to Come* (1936), Cosgrove (1998) comments on the separation of subject and object. In the cinematic version, the underground city portrays a modernist architectural environment dominated by technology and the rapid movement of vehicles. In contrast, the rural spaces of the film are designed for relaxation and aesthetic pleasure to be viewed at a distance from the window of a fast-moving machine. Such scenery is constructed 'perhaps to provide, in an ill-defined "nature", a palliative to the stresses of a mechanical life. Rarely is the countryside productive, rather it is appropriated as a distanced visual delight' (Cosgrove 1998: 263).

At the time when the film was made, the seductive appeal of the 'touring car' was being actively promoted and the modernity and speed of the machine on the open road featured strongly in advertising by the motor industry. Wright (1985: 56–68) discusses Shell's success in encouraging motorists to explore the countryside and historic towns, while associating itself with the preservation movement. From the mid-1930s, for example, Shell published a series of *County Guides* written by authors and poets with an interest in topography. In wartime, however, fuel was strictly rationed and unnecessary journeys forbidden. Nevertheless, as the country was once again threatened by invasion, guides to the nation's heritage printed on coarse utility paper sold well. In post-war, post-imperial Britain, car ownership became available to a much wider public and, with rising prosperity in the late 1950s and 1960s, the oil company's campaigns became more national in their appeal. Thus, through the novel medium of television advertising, Shell addressed the 'citizen motorist' with its *Discovering Britain* series of three-minute travelogues.

Road to nowhere?

Construction of the national motorway network in the 1960s and 1970s offered further possibilities for leisure motoring and there was a steady rise in day trips and short breaks, with visits to the countryside and heritage sites increasing still further in popularity. The rapid development of a mass market for international air travel contributed to the decline of many traditional seaside resorts, unable to compete with low-cost charter packages to the Mediterranean. On the other hand, in some remoter regions such as the West of Ireland and the Highlands of Scotland, the development of air transport and the flexibility of fly-drive holidays or air-coach itineraries boosted inbound tourism. All this led to fears of a 'leisure explosion' and concern that the centrifugal and dispersed pattern of travel would transform the landscape into a dreary uniformity. Such anxieties were, for example, articulated by the architectural critic Gordon Cullen (1971) in his comparison between the seventeenth- and twentieth-century landscapes:

> the most dramatic change is to be seen in the mobility of the individual ... The old order achieved its typical pattern of compact cities and open country because of the tediousness of travel, which forced people to gravitate to centres. Today just the opposite obtains and people cannot leave each other fast enough. We appear to be forsaking nodal points for a thinly spread, coast-to-coast continuity of people, food, power and entertainment; a universal wasteland.
>
> (Cullen 1971: 59)

Since the mid-1960s, there has been a growing awareness of environmental issues with pressure groups challenging the assumed benefits of major projects such as motorways and airports (Lowe and Goyder 1983). Perhaps reflecting these wider contemporary concerns, some British artists have returned to landscape themes, exploring relationships between the individual, society and the 'natural' environment. A notable example is Richard Long, whose conceptual art conflates the actuality of landscape with its representation. Some of his earlier work makes reference to long walks through open country illustrated by photographs. These can be interpreted as a combination of maps and chronological narrative and can be seen in work like *A Hundred Mile Walk* (1971–2). Other work has included ephemeral impressions *on* as well as *of* the landscape itself: footprints and a cross in a field of daisies marked out by picking the flowers (Spalding 1986: 221). To some viewers, however, such statements echo the early Romantics. Graham-Dixon (1996) argues that the intellectual genesis of such 'landscape art' reaches back even further to the Reformation, demonstrating a delight in God's creations and a rejection of idolatry. Long's 'sculptures', for example, include commandingly simple arrangements of stones set on their ends in a line or a circle and Graham-Dixon comments:

> It is true that the British mind has been much preoccupied with nature and the natural world during the course of the last four centuries. However, the origins of that preoccupation do not lie in painting but in the radical theology of the Reformation – not in a tradition of art, in other words but in a tradition of anti-art ... These exercises have been carried out for no apparent purpose other than to mark time spent devotedly paying homage to the natural world. Long absents himself from his work to present us with what he hopes will be an experience of renewed awareness, a small epiphany of beauty. There are religious overtones to this.
>
> (Graham-Dixon 1996: 218)

The appeal of such 'anti-art' may well reflect deeply-ingrained Protestant attitudes inherent even within a largely secular culture. The validity of distinguishing a unified, 400-year disposition of the 'British mind', however, may be disputed in a nation with a complex social history where today's population may seem more divided, its beliefs and attitudes more diverse than ever before. In the closing decades of the twentieth century, accelerated globalisation of capital and restructuring of the economy had important social ramifications. The loss of established mining, manufacturing and distribution industries brought with it the hardship of unemployment, as well as associated problems of poor housing and health, underachievement in education, and low self-esteem for many people. These factors reinforced the alienation of an 'excluded underclass', spatially concentrated in regions of industrial decline and inner city areas and about whose leisure little is known (Social Exclusion Unit 1998).

In contrast, those who prospered in the 1980s and 1990s included people who found careers in knowledge-based industries. Post-industrial economies such as those of the United Kingdom and France required the growth of a service class of educated professionals. The lifestyle choices of this section of the population have been relatively well documented, not only through market research by leisure and tourism providers but also in academic studies and analysis. Bourdieu (1984: 354–71) highlights the ascendancy of people in these knowledge-based industries such as public relations, media and communications and education. Typically, they possess a high level of cultural capital, a hunger for experimentation and a desire to learn from new experiences. Featherstone (1991: 44) emphasises the essential restlessness of people who are aware of the range of opportunities open to them. They lack the anchoring of a specific locale, refusing to be marked out by the essentialist labels of an industrial era. These 'postmodern' characteristics are expressed in their discerning tastes and footloose mobility.

Fashion, consumption and novelty have thus become important ingredients of the 'new tourism', and the media can exert a strong, if volatile, influence on demand. In some respects, the Victorian literary pilgrimage has been adapted to an age of mass communications, as people seek out the locations where their favourite films or television series were made (Ousby 1990: 187). Schofield (1996) discusses this idea in relation to a case study of the 'Hollywood of the North' experience, where the 'real' city of Manchester is interpreted with reference to its celluloid image, both past and present. After a coach tour, visitors are taken to a cinema to watch a film such as the adaptation of Greenwood's *Love on the Dole*. Unremarkable buildings, streets and districts acquire interest, status and ambience through their role as film location or association with the famous. Shields (1991) previously commented on the role of *Coronation Street* in constructing an 'identity of place' for Manchester. The phenomenon of 'teletourism' is therefore attracting increasing academic research and Tooke and Baker (1996) provide a number of examples where television programmes have stimulated substantial visitor flows.

A thriving visitor economy is now widely regarded as a desirable generator of income and employment, and place marketing has become a strategic element of planning and regeneration in areas that have suffered the decline of traditional industries (Ashworth and Voogt 1994). As a result of the frenetic desire of postmodern consumers to discover 'new' places, and the willingness of destinations to accommodate them, the tourist map now includes an eclectic range of themes and diverse locations. Destinations communicate their desirability to market segments through appropriate messages and media, and the fast-expanding World Wide Web allows the convenience of electronic 'visiting' with scope to foster 'endearment' to discerning, computer-literate travellers (Cano and Prentice 1998). As each destination strives to develop and accentuate its *genius loci*, the thesis that greater personal mobility necessarily leads to the 'flatscape' of an undifferentiated wasteland now seems hard to justify, although the depth of place-knowledge and understanding gained by the passing visitor may be open to question.

Overview

As Wallace (1993) comments, it is difficult for those of us living in comparative affluence in developed countries to dissociate the idea of travel with the prospect of fun and recreation. Nevertheless, the idea of 'travel for travel's sake' is a fairly recent notion, facilitated by developments in transport and communications: a social construct with wealth and leisure time as prerequisites for participation. Those who are prosperous and mobile may now enjoy complex lifestyles that blur the boundaries of work and non-work, liberated from dull routines and the constraints of physical space that were once accepted as the norm. Those who are poor and immobile are simply gazed upon or rendered invisible. Fast-moving vehicles, from which the passing scene is briefly glimpsed, have become icons: touring motorcycles, fast cars, high-speed trains, cruise ships and jet aircraft. Such symbols make an explicit association between pleasure and power, not only in its literal sense of exerting mechanical force, but also in its broader allusions to sexual potency, social superiority and the capacity to control.

The restless joys of travel can now be represented with spectacular realism through the moving images of cinema, television, video and multimedia. Feifer (1985) discusses the McLuhanesque global village, a network of communications from which it is possible to gain a little knowledge of a great many places and times. Such landscape images are consumed, free from the context of any universally held world view. For some years, then, it has been possible to experience the passive seeing function of tourism without the need to travel. Now, as microelectronics-based telecommunications merge with digital computer and media developments, the barriers of space-time are transcended with even greater ease. Multimedia technology offers the prospect of simulated, and even virtual travel, a 'vehicle' which permits us to roam around the world and accumulate tourist experience without ever boarding a mechanism of physical transport (Rojek 1998).

The paradox of the armchair itinerary is taken a stage further as the illusion of travelling through distant landscapes, and even backwards or forwards in time, is enjoyed as home entertainment. The new or post-(mass)tourist is, however, far more than a passive consumer of images. He or she derives pleasure from comparing image with reality, self-consciously playing the role of outsider-tourist (Urry 1990). As Richard Long's conceptual art suggests, the actuality of landscape is to be anticipated, encountered and then relived through reminiscence. Developments in communications and media technology have seldom dampened the demand for travel. Indeed, for those who enjoy the privilege of social and physical freedom of movement, such technologies seem to offer stimulation by heightening awareness of the possibilities and creating the desire to experience places and pleasures of the journey at first hand.

4 Valuing the countryside
Leisure, tourism and the rural landscape

Introduction

It seems no coincidence that the first nation to experience industrialisation and mass migration from rural to urban areas should have developed such intense nostalgia for the countryside. For at least six generations the majority of Britain's population has lived in towns and cities. The desire for outdoor recreation in more open, greener surroundings is, perhaps, an understandable response to the confinement and pressures of living in an urban environment. Nevertheless, it is hard to explain the fetishism for particular landscapes, or the expectation of conformity to 'acceptable' patterns of behaviour by visitors to the countryside, except by reference to the hegemony of a dominant culture that has idealised the notion, if not the harsh reality, of rural life. Through visual imagery, words and music, the wholesome appeal of the countryside has thus been celebrated and presented to urban visitors, nurtured and reinforced through guide-book culture and deliberate place promotion. As Bunce (1994) has argued, the countryside ideal can be interpreted as a social construct that emerged and evolved along with industrial society.

As rural Britain engaged with a fully capitalist system of production, distribution and exchange, the economy of the countryside was transformed. Through the later eighteenth and nineteenth centuries, agricultural production was intensified and mechanised to supply expanding urban markets. Even the most remote regions were connected to the new arteries of turnpike roads and later to a trunk railway system. A parish ceased to be the centre of the area its inhabitants could survey. Increasingly, it became part of a network of routes with no obvious centre but the metropolis (Barrell 1972: 86). From the perspective of the rising urban middle classes, distant regions, that had once seemed inaccessible, could now be enjoyed as leisure and tourism destinations. To this audience of actual or would-be visitors, such rural landscapes could evoke the fascination of a lost domain, somehow 'left behind' in the modern race for progress, but giving rise to comforting images of 'the homely, the stable, the ahistorical' (Bermingham 1986: 9; Shields 1991: 3). In this context, 'unspoilt' and 'natural' countryside acquired a special mystique.

The allure of such landscapes has been a significant factor in the spatial development of leisure and tourism in the British Isles. Canons of taste derived from Augustan and Romantic culture have thus determined the selection of sights we seek out today as major features of the tourist map (Ousby 1990: 9). The idealisation of particular forms of scenic beauty, experienced by the lone wanderer seeking refuge from the demands of urban society, was also to inspire many of those who championed the public's right of access and protection of valued rural landscapes. Such campaigns, in their turn, shaped public policy, including the designation of National Parks and the desire to accommodate visitors in particular areas and steer them away from others. It would be overly simplistic, however, to assume a straightforward diffusion and assimilation of attitudes once shared by a small coterie of artists and intellectuals. Closer consideration reveals a rich plurality of traditions, activities and landscape preferences that our dominant conceptualisation of the countryside has tended to obscure.

This chapter examines the idealisation of rural Britain as a place of retreat and refreshment for the benefit of urban visitors. It discusses the enthusiasm for the countryside and various forms of outdoor recreation demonstrated by diverse sections of urban society from the early nineteenth century, and their response when excluded by the rural establishment. The pursuit of scientific and artistic study, as well as self-improvement through countryside recreation, created a mosaic of clubs and associations some of which formed alliances to make the case for public access and protection of valued landscapes. In the post-war period these principles were enshrined in legislation, but implementation of policy to encourage access to 'a countryside for all' proved problematic. Over the past decade or so, rural destinations, especially those where the agricultural sector is in decline, have been actively developed and promoted by place marketeers. The presentation of the countryside as a landscape of leisure and tourism consumption has significant implications, not only for host communities in rural areas, but also for social equity in the provision of outdoor recreation.

Nature was his book

Raymond Williams (1973: 1) emphasises the accretions of meaning that have developed around the idea of 'the country' in antithesis to 'the city'. The positive attributes of the country include the natural way of life, peace, innocence and simple virtue. Negative ones include backwardness, ignorance and limitation. The Arcadian pastoral convention that country dwellers live a happier and morally just existence gained prominence in English literature during the eighteenth century, but as Thomas (1983) comments, the notion of the contented shepherd or swain leading a blameless and independent life had no basis in social reality. Agriculture was among the most ruthlessly developed sectors of the economy, small live-stock farmers were declining, wage labour was common and the vices of avarice, oppression and hypocrisy were at least as prominent in the countryside as in the town. Furthermore, Nature 'herself' seemed in retreat as landowners exploited new scientific developments in cultivation and husbandry to raise the output of agriculture. Parliament was liberally approving the enclosure of wastes and commons, and walkers of all social classes encountered thorn hedges and other new physical barriers that restricted their freedom to roam at will.

For those who worked on the land, agrarian change marked the end of customary obligations that had cushioned life against the seasonality and uncertainties of rural employment. Villagers could no longer graze their livestock on common land, and opportunities to take game and gather winter fuel were withdrawn. Draconian game laws were introduced, with cruel punishments for those convicted of poaching. Landed interests were further consolidated by the 1815 Stopping-Up of Unnecessary Roads Act which simplified the closure of established rights of way. In former times the commons, wastes and footpaths had played an important role in the social life of the countryside. In some localities access was now denied and people's enjoyment of traditional sports and outdoor recreation restricted (Donnelly 1986). The customs that had characterised a more paternalistic social order were thus eroded. Increasingly, 'the poor' were regarded by those with wealth as an undifferentiated class whose presence could arouse fear as well as charitable concern. With the transition to a capitalist economy their condition was reduced to that of a landless proletariat. As a result, the labouring poor in town or country became more dependent on their employers and more tractable to their discipline (Barrell 1983: 3–4).

The hardship experienced by many of the rural poor was reflected in the poetry of former farm labourer John Clare, in reference to his native Cambridgeshire. Williams (1973: 134–41) acknowledges Clare as a deeply significant figure, not only for his poetic innovation, but also for the way in which he internalised the far-reaching social changes of his time. In 'The Village Minstrel' (1821), their outward expression in the landscape is chronicled:

> There once were lanes in nature's freedom dropt,
> There once were paths that every valley wound -
> Inclosure came, and every path was stopt;
> Each tyrant fix't his sign where paths were found,
> To hint a trespass now who cross'd the ground:
> Justice is made to speak as they command . . .

As Clare (1821) notes, the liberty of England was being suppressed and its rural workforce impoverished and subjugated:

And every village owns its tyrants now,
And parish-slaves must live as parish kings allow.

(cited in Williams 1973: 136-7)

To the wealthier town-dweller, however, the countryside was regarded as a desirable place of retreat. As urban life, particularly in London, became less tolerable through its squalor, disease, pollution, noise and bad odours, a privileged minority took to their summer houses in nearby countryside for relaxation and refreshment. Travel to enjoy peace and fresh air at weekends became easier, with an increasing use of private carriages from the late sixteenth century. Two centuries later, the pleasures of a rural retreat reached the very height of a fashion associated with an Arcadian rustic ideal and exemplified by Queen Charlotte's *cottage ornée* built for her at Kew in 1772. Over the second half of the eighteenth century there were also significant shifts in landscape taste which greatly influenced the notion of what constituted attractive countryside. Hitherto, the straight lines and neat symmetry of cultivation had appealed to the eye of the educated observer. Now, the cultured elite was acquiring a new taste for rough and rugged scenery.

The development of new farming techniques was recorded in detail and with delight by apologists such as Arthur Young and the Board of Agriculture. Nevertheless, as Prince (1988: 98) has pointed out, their physical manifestation in the landscape was ignored by most contemporary artists. This was generally the case even during the period of greatest change and in paintings commissioned by landowners that had profited most from agrarian improvement. Seldom do the newly enclosed fields, farmsteads, reclaimed heaths or recently introduced crops feature as fitting subjects. Instead, landscape scenes of the earlier period refer back to classical antiquity. These give way to Picturesque views that 'concentrate on crags, cliffs, waterfalls, woods, dead trees and crumbling ruins', and Romantic pictures that 'seek to arouse powerful feelings of wonder and awe at the beauties and sublimities of cloud forms, geological structures and plant life. All these styles of painting seem to treat features of agricultural change as intrusions' (Prince 1988: 98).

As the next chapter illustrates, the 'discovery' of the British countryside in the later eighteenth and nineteenth century owes much to innovative painters, writers and composers whose art reflected a deeper humanistic philosophy. The studied distancing of subject from observer, and pedantic atomisation of the view characteristic of Picturesque taste, yielded to a new sensibility. For the Romantics, love of the countryside involved direct experience and immersion in Nature. Towards the end of the eighteenth century, the practice of walking long distances was liberated from its former connotations of need, homelessness and suspicion of criminality. Among the leisured elite, people began to think of walking through open country as a pleasure: satisfaction of impulses towards the re-creation of self, reconnection with nature and so with the divine, and continuity of sense, mind and spirit, community and communal past (Wallace 1993).

In the early nineteenth century, the mass-production of literary and artistic images was made possible by the development of subscription and then by commercial publication of the modern kind (Williams 1961: 50). Eulogies to the joys of walking through wild and beautiful landscapes thus became familiar to the comfortable and literate middle classes. In prose, the soothing emotion associated with everyday features of the rural landscape was expressed in William Hazlitt's essay 'On the Love of the Country' (1814) published in *The Examiner*:

> the most insignificant and rudest objects are often found connected with the strongest emotions; we become attached to the most common and familiar images as to the face of a friend whom we have long known, and from whom we have received many benefits.
>
> (Hazlitt 1814)

As Wallace (1993: 176–7) comments, Hazlitt walks to abstract himself from the material and social demands of present perception, disengaging from adult social intercourse to regain the receptivity of childhood. Reference is made to happy and nostalgic recollections of the countryside drawing upon the author's childhood and connecting the present with the past. Enjoyment of the country could therefore be seen as a highly personal and reflective experience. This process of renewal required withdrawal from the restrictions of urban life and a temporary detachment from society. Hazlitt expands on this theme in 'On Going a Journey' (1822) in *New Monthly Magazine*:

> One of the pleasantest things in the world is going a journey; but I like to go by myself. I can enjoy society in a room; but out of doors, nature is company enough for me. I am then never less alone than when alone.
>
> 'The fields his study, nature was his book.'
>
> I cannot see the wit of walking and talking at the same time. When I am in the country I wish to vegetate like the country. I am not for criticising hedgerows and black cattle. I go out of town in order to forget the town and all that is in it. There are those who for this purpose go to watering places, and carry the metropolis with them. I like more elbow room and fewer encumbrances.
>
> (Hazlitt 1822)

By the early nineteenth century, when Hazlitt expressed these sentiments, the landscape of many counties was somewhat at odds with the notion of an unchanged Nature. As the process of agrarian improvement and industrialisation ran its course, devotees of the new creed took solace in the wilder scenery of more remote regions. Set against the development of a capitalist society, Romantic ideology elevated the idea of the sublime, which was to be found in the soul of the individual especially the poet or artist. In the external world the sublime could be found in those places that emphasised the insignificance of humankind: barren mountain recesses, storms, seas and the night (Cosgrove 1998: 230–1). Pioneer hillwalkers sought such an experience of landscape, celebrating spontaneity and the lack of formal rules. When Wordsworth first published his *Guide Through the District of the Lakes* (1810), the poet hoped to bring others to a new appreciation and understanding of the countryside, encouraging 'habits more exact and considerate observation than have hitherto been applied to local scenery' (Moir 1964: 155). Describing the natural beauty of the Lake District, he argued for the right to roam freely in the region. The Lake District should thus be deemed 'a sort of national property in which every man has a right and interest who has an eye to perceive and a heart to enjoy' (quoted in Glyptis 1991: 26). The Romantic ideal was to be deeply influential in formulating the principles of public access to the countryside and selection of landscapes deemed worthy of protection. Nevertheless, the social and cultural origins of the 'outdoor movement' were remarkably diverse.

Access and exclusion

From the early nineteenth century, the desire for outdoor recreation was expressed by some of the poorest among the industrial proletariat. As yet, the majority was illiterate, their opinions unrecorded. Nevertheless, the fragmentary evidence of observers suggests that people from manufac-turing areas such as Bolton, Burnley and Oldham were making extensive use of the established network of footpaths to enjoy fresh air and exer-cise in nearby countryside during their limited free time (Stephenson 1989: 59). Unfortunately, in Lancashire and other counties, estate owners were applying the Stopping-Up Act with the specific aim of resisting such pres-sures, and this practice aroused concern over the health and well-being of industrial workers (H. Taylor 1997: 24). With varying degrees of success, sympathetic middle-class reformers and radicals established asso-ciations to oppose further footpath closures or loss of open space, early examples being those for the areas around York (1824) and Manchester (1826). Questions were raised in parliament, a select committee investi-gated, and reforms followed. The amenity value of ancient footpaths as public walks was thus acknowledged.

From the 1830s, the railways presented novel possibilities for visits to more distant countryside, as well as the coast and urban centres. Rail excursions, sometimes supported by charitable benefactors, enabled people of modest income to travel *en masse*. There were, nevertheless,

authoritative voices who opposed such freedom of movement by the proletariat. Tourism was entering a new phase, and the Romantics' idealisation
of solitary contemplation of the wild and sublime seemed to sow the seeds
of its own destruction. Now, the social tone might be lowered still further
and attractive landscapes transformed into vulgar inland resorts.
Subscribing to this view, Wordsworth (1844) supported the campaign
against the proposed branch line to Lake Windermere that threatened an
invasion by 'the whole of Lancashire and no small part of Yorkshire'. In
his opinion, appreciation of a fine landscape was not innate. Rather, it
required an education in aesthetics. In his letter to the *Morning Post*
(1844) the ageing poet railed against a mass influx of 'artisans, labourers
and the humbler classes of shopkeepers' who should not be tempted to
'ramble at a distance'. Instead, they should be encouraged to:

> make little excursions with their wives and children among the neigh-
> bouring fields within reach of their own urban dwellings whither the
> whole family might stroll, or be conveyed at much less cost than
> would be required to take a single individual to the shores of
> Windermere by the cheapest conveyance.
>
> (Walvin 1978: 23)

There seems little evidence, however, that the countryside sustained much
damage. Walvin (1978: 23) describes an outing to the small village of
Littleborough on the Manchester–Leeds line in September 1839:

> They were mostly clerks, warehousemen, and the decenter sort of
> operatives, with some females and children ... The day passed off
> without the least irregularity or disturbance, and not a small propor-
> tion of the visitors attended service in the afternoon at the church.
>
> (Stephenson 1989: 58)

Furthermore, among the working-class communities to whom Wordsworth
referred, the desire for self-improvement of mind, body and soul was an
important stimulus. To the leadership of Non-conformist chapels, countryside recreation offered a respectable alternative to more hedonistic and
corrupting pleasures such as going to the races. A leading and energetic
spirit in this movement was the Reverend Thomas Leonard, a Congreg-
ational minister who later resigned his ministry to become secretary of
the Co-operative Holidays Association. He subsequently helped to found
the Holiday Fellowship, the Youth Hostels Association and the Ramblers'
Association. A more secular, but no less purposeful agent of progress and
self-improvement was the socialist Clarion rambling and cycling clubs that
established many local groups through to the First World War.

   The Mechanics' Institute, a forerunner of the further education college,
also played an important part in the drive for self-improvement, an

outcome of which was the formation of rambling clubs. With growing literacy among the industrial workforce, these were established to help industrial workers enhance their skills through night classes. By the mid-nineteenth century, some had developed curricula that encouraged liberal study, including the natural sciences. Field visits to the surrounding countryside became an important feature of their teaching methods. The institutes produced some inspiring tutors, many of whom had themselves worked in factories or mines since childhood (Stephenson 1989: 66–7). In 1848, G. S. Phillips of the Huddersfield Institute published a guide-book written to introduce those confined 'day after day in closed workshops' to the beauty and grandeur of the nearby countryside. He argued that such beauty should be regarded as 'an inalienable common right; and he who carries the true-seeing eyes in his head, no matter how poor he may otherwise be, is the legitimate lord of the landscape' (quoted in Stephenson 1989: 66).

Among the Victorian middle classes there was also a keen demand for active recreation in the countryside. Some took up the 'traditional' country sports such as hunting, fishing and shooting, but many others took part in a range of pursuits that were essentially urban-based. Characteristically, participants were educated men of the professional classes who followed their mutual interests through clubs and societies. To some extent, these followed in the tradition of the gentleman dilettante, with an active interest in scientific pursuits such as botany, geology, ornithology and archae-ology. In the case of mountaineering and hillwalking clubs, there was also a strong emphasis on the individual's powers of endurance and ability to compete with the elements. In Scotland in 1891, for example, Sir Charles Munro published a list of 283 peaks over 3,000 feet, establishing the famous challenge to climb them all. There were also middle-class groups who participated in rambling as an activity to promote good health and companionship, demonstrating its attraction as a sociable as well as a solitary activity. Perhaps the most famous was Sir Leslie Stephen's 'Sunday Tramps', a literary and intellectual circle, based in London and the home counties from 1879 to 1894. Stephen, who was also a keen mountaineer, thus managed to combine vigorous long distance walking with an appre-ciation of Nature and high-minded discussion (Taylor 1997: 60–6).

By the end of the nineteenth century, there was a certain unity of purpose within the developing outdoor movement. A particular issue that strengthened its resolve was loss of access to the open countryside of moors and mountains as large tracts of land were taken in for commer-cial hunting and shooting. In Northern England and Scotland, sheep pasture and other farmland was being covered as the value of land for game exceeded that which could be obtained from agriculture, especially from the 1870s (Donnelly 1986: 217). In Scotland, public exclusion by landowners challenged a deeply felt traditional right to cross open moors and mountains. The argument for access aroused patriotic sentiment as estates were then being acquired by English landed gentry. In response to the injustice of these access restrictions, James Bryce, Liberal MP and keen mountaineer, introduced the Access to Mountains (Scotland) Bill, representing the first attempt to legislate a general right to roam. In 1884, this Bill proposed that 'no owner of uncultivated mountain or moorland should be entitled to exclude any person from walking on such lands for purposes of recreation or scientific or artistic study'. The Bill was, nevertheless, withdrawn, as were a further eleven unsuccessful attempts by Bryce. For over half a century, the opposition of landed interests in both Houses was to remain an obstacle to legislative reform.

Voluntary action to acquire land and property for the public good met with more immediate success than compulsion through legislation. The National Trust for Places of Historic Interest and Natural Beauty was established in 1895 as part of wider paternalistic concern to preserve open space and heritage sites. The Trust was founded by Octavia Hill, a prominent and influential social reformer who had been a prime mover in the Commons, Open Spaces and Footpaths Preservation Society and its action to protect remaining unenclosed land. The leadership of the Trust included other eminent figures, notably John Ruskin (Hill's former tutor) and Pre-Raphaelite painter Holman Hunt. The organisation's mission was not only to acquire sites and hold them in trust for the nation, the aesthetic of landscape, too, was an important concern of its founders, and the Trust soon threw its weight behind the preservation cause at national and local level. Its early campaigns included opposition to the building of the Snowdon mountain railway, encroachment on Hampstead Heath and disfigurement of the countryside by telegraph poles (Lowe and Goyder 1983: 138).

Landscape fit for heroes

The early part of the twentieth century was a period of rapid social change and there was an expectation of reform, especially after the First World War. In this climate, there was a significant expansion and broadening of participation in countryside recreation, and pressure was stepped up to secure improved public access. Increasingly, an alliance of organisations concerned with outdoor recreation, and built on the socially disparate foundations discussed above, became a movement conscious of itself. A diverse range of leisure uses of the countryside had emerged, some of which were mutually incompatible, but a cause which had already united them was their unwillingness to be beholden to owners of country estates. In practice, recreation-seekers frequently encountered obstructions that restricted their enjoyment. As H. Taylor (1997: 2) has argued, there was also an underlying unifying ideology 'situated within the wider context of notions of "rational leisure" as a self-improving and character-building alternative to traditional forms of popular recreation'. As with the contemporary heritage movement, another common theme was the rejection of mercenary values, and revulsion at the commercial exploitation of the countryside they cherished.

It was not uncommon for defence of valued countryside to be expressed in stirring patriotic terms. Disfigurement of landscape by 'spectacular advertisements' was a case in point. J. Taylor (1990) discusses the Society for Checking the Abuse of Public Advertising and its successful campaign for legal controls to protect amenity on similar lines to by-laws for noise or smoke abatement. The organisation's journal argued:

> It is absurd to talk of English freedom and prosperity and the greatness of our imperial mission, if no regard be paid to the beauty of the landscape, to dignity and propriety in the common round. We are proud of our fatherland, we care for our countrymen, and wish to make the national domain pleasant and worthy of a great people
> (quoted in J. Taylor 1990: 191)

With little or no planning control, areas of scenic beauty were also threatened by speculative building. Low-density suburban development was proceeding at a fast pace, with the availability of mortgages, commuter transport and motorcars which middle-class people could afford. With a serious depression in agriculture, owners of suitably located farmland were inclined to sell. The influential Council for the Preservation (now Protection) of Rural England was set up in 1926 to counter this threat, as well as other forms of urban 'encroachment' on the countryside. The National Parks Committee of Inquiry (1931) emphasised a growing national concern:

> there are in this country areas of peculiar interest to the nation as a whole – typical stretches of coastline, mountain regions, moor and downs, riverbanks and fen. These areas constitute an important national asset and the Nation cannot afford to take a risk that they will be destroyed or subjected to disorderly development.
> (quoted in Cherry 1985: 130)

Meanwhile, there was a widespread enthusiasm for physical fitness, fresh air and sunshine. Hiking, rambling and cycling had gained enormous popularity with working people and those experiencing the enforced 'leisure' of unemployment. Now, even the poorest urban dwellers could reach good hiking country by bicycle or public transport, and many were keen to roam. Motor coaches offered scheduled services and ran excursions facilitating visits further afield (Gruffudd 1994: 248). By the early 1930s there were an estimated half a million regular walkers and rambling had become a mass leisure activity (Blunden and Curry 1989: 30). A Chinese visitor to Britain, a landscape artist named Chiang Yee (1937:

4–5), observed this rather strange phenomenon. Like Hazlitt, he had a preference for solitude and silent contemplation. When touring North Wales, he gazed upon Snowdon, Swallow Falls and other features which struck him as 'an entirely commonplace scene' had it been along the south bank of the Yangtze River. As he was in the company of a party of hikers, however, he felt obliged to follow them:

> it left me little time to look at the surroundings as we were walking. All members of the party were very kind to me and from my lack of conversation they seemed to conclude that I was lonely, so they frequently came to me and offered help with the historical background or some legends and stories surrounding this spot or the other. This kind of explanation continued unceasingly as we were on the march. (We were more than forty, and as we walked through the passes we really seemed like part of an army).
>
> (Chiang Yee 1937: 4–5)

Rambling clubs proliferated in number and membership throughout the country, and regional federations were formed to develop a consensus on issues relating to access, as well as 'uglification' of the countryside. A practical and popular outcome of a special meeting of the Liverpool and District Federation in 1929 was the Youth Hostels Association, formed to widen opportunities for young hikers and others using a network of nearly 400 hostels throughout the United Kingdom by the late 1930s. Legal rights were still a long way off, but as the democratising spirit of the 'movement' gained confidence, some clubs and other pressure groups campaigned for access where landowners excluded them. Some favoured direct acts of defiance, notably the 'mass trespass' over the Duke of Devonshire's land at Kinder Scout in 1932 that can be seen as an explicit expression of class struggle directed against the landowner. The subsequent arrest and imprisonment of its organisers gained much publicity, but also the disapproval of those who favoured consensus-building and parliamentary reform. The latter included leaders of the fledgling federation that in 1935 became the Ramblers' Association, a national organisation whose primary aim was legislation to secure access to moors and mountains.

To some observers, however, the behaviour of hikers was not always appropriate, and misgivings were expressed. Like Wordsworth almost a century before, there were those who felt that visitors needed prior tutoring. For example, in 'The people's claim', C. E. M. Joad (1938: 73–4) referred to 'the interest of the people in the English countryside and their consequent claim upon it'. Nevertheless, he cited a number of examples of how 'the townsman outrages the country' through misbehaviour. It was a demonstrable fact that they were:

not yet ready or able to exercise their rights without damaging that
to which they lay claim … there are the hordes of hikers cackling
insanely in the woods, or singing raucous songs as they walk arm in
arm at midnight down the quiet village street. There are people, wherever there is water, upon sea shores or upon river banks, lying in
every attitude of undressed and inelegant squalor, grilling themselves,
for all the world as if they were steaks, in the sun … there are fat
girls in shorts, youths in gaudy ties and plus-fours, and a roadhouse
round every corner and a cafe on top of every hill for their accommodation.

(Joad 1938: 73–4)

More generously, Patrick Abercrombie (1938) advocated legislation for
countryside planning to enshrine 'the use of the open country as the
greatest of all parks'. A new contract between town and country was
needed. This must acknowledge that 'it is not only the public open spaces
but the fully used farm land with its oldest of human occupations which
is so refreshing to the townsman' (Abercrombie 1938: 138).

A people's charter for the open air

Some fifty years after Bryce's first attempt, another abortive Access to the
Mountains Bill (1939), brought bitter disappointment. But, in 1945 the
incoming Labour government's manifesto included comprehensive land use
planning and better access to the countryside. Furthermore, there was the
enthusiastic commitment of Lewis Silkin, the first Minister of Town and
Country Planning, and a keen rambler. Following the Report of the
National Parks Committee (1947), the 1949 National Parks and Access to
the Countryside Act proclaimed a new era in which public access for outdoor recreation *and* protection of special landscapes were seen as complementary goals of public policy. Introducing the Bill, Silkin called it:

a people's charter for the open air, for the hikers and the ramblers,
for everyone who loves to get out into the open air and enjoy the
countryside. Without it they are fettered, deprived of their powers of
access and facilities needed to make holidays enjoyable. With it, the
countryside is theirs to preserve, to cherish, to enjoy and to make
their own.

(Hansard, 31 March 1949)

In its attempt to marry landscape protection and provision for outdoor
recreation, the development of public policy for national parks in Britain
illustrates a process of interaction characterised by the relative strengths
and the tactics of the key actors, rather than the logic or force of a

particular cause (Cherry 1985). A significant outcome of this approach was the selection and designation of ten national parks in the period up to 1957, while policy for recreation in countryside elsewhere remained largely unstated. During the war years, civil servant John Dower, with a strong predilection for mountain and moorland scenery, had persuasively restated the case for national parks and mapped out areas he considered appropriate. The 1949 Act enshrined the idea that national parks should be chosen for their natural beauty with regard to landscape character and proximity to centres of population. As Shoard (1982: 65) points out, however, most are a long way from major cities, including London, Southampton, Birmingham and Bristol. Some explanation of this bias for wilder scenery, a preference not necessarily shared by the majority of the population, lies in the personal taste and enthusiasm of a small group of people who shaped public policy.

Another compelling reason for omitting lowland scenery such as chalk downland in the designation of National Parks, was the priority given by government to maximising food production and the desire to be less dependent on imports. British agriculture was now protected and supported by public finance, technical advice and a relaxed planning regime. A significant increase in crop yields and livestock production was made possible by new farming technology. With the use of large-scale equipment for ploughing and harvesting, hedgerows and small patches of woodland came to be regarded as obstructions. Winding streams were culverted, marshes drained and downs and moors converted to arable land. Some areas experienced the most dramatic landscape change since the agrarian improvements of the eighteenth and early nineteenth centuries. The countryside was far more prosperous than in pre-war days but, to many urban visitors, landscapes were less attractive, and in some areas there were increased problems with access.

Meanwhile, there were important shifts in the patterns of personal mobility compared with pre-war days, when the case for access and landscape protection had first been made. In particular, the assumption that most visitors to national parks and the wider countryside would continue to arrive by train, bus or bicycle was shown to be ill-founded. In the USA, Henry Ford had mass produced his 'touring car' since the 1920s. On both sides of the Atlantic, attractive leisure motoring maps and guides helped promote the freedom of the open road. Gold and Gold (1995) describe the re-establishment of tourism in Scotland after six years of war:

> This was frontier country: lonely, wild and grand. The roads may have been narrow and the accommodation scarce, but there were few cars on the road and few restrictions on where they could go or park. The caravan, and the car which towed it, were powerful symbols of freedom and choice.
>
> (Gold and Gold 1995: 118)

The motorist had the exciting privilege of independence from railway routes and timetables resulting in a renewed sense of discovery. In post-war Britain, the pattern of car ownership followed the same rapid growth curve that the USA had experienced. In 1951, less than 15 per cent of British households owned a car, but this rose to 31 per cent in 1961, and to 52 per cent in 1971 (Department of Environment Transport and the Regions 1998). It is understandable that those who had recently acquired cars should exercise their new-found independence. Yet, unlike North America, the intricate pattern of narrow, winding roads and ancient settlements was to prove ill-suited to motorised traffic in many rural areas. The huge growth in recreational trips, associated with increased affluence, leisure time, and the rapid expansion of private motoring, once again stirred outrage against the invasion of 'our countryside'.

By the 1960s, public policy seemed more concerned with controlling the flow of recreation-seekers who were invading the countryside. An influential paper by M. Dower (1965) for the Civic Trust argued the case for leisure planning and visitor management. In this, he raised the spectre of a multitude swarming out of cities 'like ants' scurrying from coast to coast. Somewhat at odds with Silkin's post-war vision, emphasis was placed on protecting the urban fringe areas of countryside most accessible to population centres. The moors and mountains would therefore be preserved for the true wilderness-seeker, concentrating the crowds into places that could accommodate them and keeping the high, wild places for those seeking solitude. The White Paper *Leisure in the Countryside* (1966) led, two years later, to the establishment of the Countryside Commission (now Agency), with a remit to facilitate conservation and enjoyment in the wider countryside in England. It also outlined the principle of country parks. These, along with picnic sites, would 'ease pressure on more remote and solitary places', and 'reduce the risk of damage to the countryside, aesthetic as well as physical which often comes when people settle down for half an hour or a day when it suits them "somewhere in the countryside" to the inconvenience and expense of those who live and work in the locality' (Minister of Land and Natural Resources and Secretary of State for Wales 1966).

It was becoming fashionable to highlight contradictions within the 1949 Act, especially conflicts between public access, the utility of farmland and the protection of landscape value and wildlife habitat. Consultants Westmacott and Worthington undertook a study for the Countryside Commission (1974) to find out 'how agricultural improvement can be carried out efficiently but in such a way as to create new landscapes no less interesting than those destroyed in the process'. The visual appeal of landscape proved, however, a variable and illusive quality which defied objective measurement: 'Unfortunately, vivid impressions of the countryside during childhood often dominate preferences in adults and can lead to a sentimental and preservationist approach to landscape conservation'

(Countryside Commission 1974: 1). The demands made upon the land-scape reflected both individual and collective needs. The balance between the various functions was continually changing and was exemplified by the rapid increase in demand for informal recreation: 'While agricultural operations, some recreational activities and wildlife conservation coexisted with little conflict until recently, the increasing intensity of both agriculture and recreation now threatens their coexistence on the same land' (Countryside Commission 1974: 1).

The idea of creating country parks and picnic sites was favoured by many local authorities and encouraged by the Countryside Commission through grant aid, the idea being to intercept outbound weekend motorists and entice them into specially designed 'honeypot' sites. Today, it is generally accepted that such facilities provide welcome opportunities, especially for parents and children to relax and enjoy themselves in the fresh air without having to travel too far from home. They offer convenience, safety and the facilities of an urban municipal park in a pleasant rural setting. Nevertheless, the tone of the rationale now seems somewhat patronising and the solutions high-handed. Why should visitors be denied the rich variety of leisure opportunities which freedom of the wider countryside can offer? Furthermore, these managed sites with off-road parking spaces were essentially designed to accommodate the car-borne visitor rather than those dependent on public transport. Two decades on, Shoard (1987: 385–96) and others were to draw attention to the condescending image of visitors to the countryside as a plague of noxious, greedy insects that had to be contained.

Despite the concern of policy-makers over the forecast explosion in demand for recreation in the countryside, the post-war vision of a people's charter was never wholly abandoned. Fears that rural areas would be overwhelmed by townspeople diminished in the wake of the oil crisis and economic slow-down of the mid-1970s. Some local authorities continued to emphasise the countryside as a community resource, but were mindful that many urban residents without cars now had difficulty reaching nearby countryside. Following the 'reshaping' of the railways, much of rural Britain was without a local rail service, and the residual bus network was in a downward spiral of falling passenger numbers, increasing fares and service reductions. In response, around half the counties of England and Wales supported or promoted recreation transport initiatives from the mid-1970s to mid-1980s. Most of these schemes were justified by enlightened concern for those dependent on public transport, and the concept received encouragement from the Countryside Commission as well as the Regional Councils for Sport and Recreation. Nevertheless, there were lessons to be learned about effective social marketing. A review of such initiatives by researchers at Manchester University found that around one-sixth had been abandoned by the mid-1980s, and half failed to meet their providers' expectations (Blunden and Curry 1988: 145–6).

Operation Gateway in the East Midlands was one such scheme that highlighted social and cultural issues regarding use of the countryside. The 'Sherwood Forester', a low-fare leisure bus network from Nottingham city centre, had been initiated by the county council, but usage was disappointingly low. Site surveys at the country parks, Sherwood Forest and other visitor centres which it served showed 'a disturbingly familiar pattern – that the largest uptake of these publicly funded facilities was from the most privileged, most mobile sections of the community' (Countryside Commission 1989: 1). To investigate, small group research was carried out with non-car-users, single parents and low-income residents of inner Nottingham and its outlying estates. The research revealed that people had many reasons, apart from lack of money, for not taking trips to the countryside. Many had heard of Sherwood Forest through Robin Hood films, but few knew where it was, how to get there or what was there. Most lacked confidence in rural bus services and few were aware of the Sherwood Forester. Timetables and maps, if available, were hard to interpret and those travelling with children were understandably worried about long, uncertain waits. Operation Gateway subsequently developed a range of outreach projects to encourage neighbourhoods and community groups to organise outings, establishing links with farmers and other hosts. Nevertheless, the process was much slower, more labour-intensive and required longer lead times than had been anticipated.

A countryside for all

By the mid-1980s there was a rising tide of criticism which questioned the wisdom of directing the majority of recreation-seekers to managed sites and facilities, while designating 'spectacular' areas of mountain and moorland scenery for 'serious' countryside users. Research raising such doubts included a seminal paper by Harrison, Limb and Burgess (1986) based on small group discussions with people in the London Borough of Greenwich, including working-class white and Asian residents. The study highlighted the value of unspectacular and unmanaged green spaces within easy reach of urban residents whose main enjoyment was simply being out and about in the countryside, having contact with the living world and undertaking active, spontaneous and 'naive' encounters. In all the groups, people frequently described the countryside as a kaleidoscope of sounds, smells, sights and tactile experiences of nature. For some, especially those who had grown up in rural areas, memories of childhood permeated the very meaning of 'countryside'. For example, Asian women with memories of the Punjab, Kashmir, Gujurat and Kenya had deep attachments to their home environments, though most had lived in Britain for at least fifteen years. One woman summarised the group view: 'For us the countryside is small lanes, going through really windy roads and probably miles and miles of fields and open spaces like this really. There's villages' (quoted in Harrison, Limb and Burgess 1986).

The small group discussion emphasised the intensity of people's experiences with nature and landscape, releasing parts of themselves normally forgotten or pushed out of sight in daily life. The study concluded that experience of the countryside was neither the passive reaction that some questionnaire surveys suggest nor a form of landscape appreciation exclusive to a sensitive elite.

Another influential study, carried out for the Countryside Commission and Sports Council (1986) further highlighted issues of social equity. In *Access to the Countryside for Recreation and Sport*, the authors drew a useful distinction between the concept of *access* – the legal right of entry – and that of *accessibility*. The latter refers to the extent to which such rights can be exercised in a particular time and place. Rights of entry are a necessary precondition, but cannot in themselves guarantee accessibility. The notion of accessibility highlights the need to understand the variety of meanings which different social groups associate with the idea of countryside. Conflicts over access issues are, therefore, very often concerned with conflicting perceptions of the countryside and what constitutes 'legitimate' recreational and sporting activity. The report of the study argued that the more mobile and better-off regular users of the countryside were probably the main beneficiaries of public policies. Among the infrequent participants, there was clearly scope to 'deepen' the experience, quality and satisfaction of sport and informal recreation participation as well as to increase its frequency. Furthermore, there were significant minorities who never visited the countryside. National programmes could be sponsored to stimulate and aid their participation.

Following a far-reaching and comprehensive review of policy, the Commission seemed to acquiesce to criticism of past practice, and to embrace a renewed concern to broaden participation. In the context of reduced subsidies to agriculture and the need to encourage diversification of the rural economy, *Policies for Enjoying the Countryside* (Countryside Commission 1987) heralded a marked change of tone:

> In the 1960s and 70s, the public 'invasion' of the countryside was interpreted as a threat. Today, with many agricultural products in surplus, the patterns of British agriculture and rural enterprise will have to respond to significant changes in demand. In this light, public interest in enjoying the countryside becomes an opportunity.
>
> (Countryside Commission 1987: 8)

The Commission's 'people-related policies' stressed the information, confidence and understanding that people require to make full use of the countryside. It emphasised the desirability of attracting occasional users of the countryside and encouraging those who visit rarely or not at all.

The latter segment represented a quarter of the population, and included 'the elderly, those on low incomes, the unemployed and ethnic minorities' (Countryside Commission 1987: 9). Its place-related policies highlighted the significance of the rights of way network and acknowledged that country parks and other managed sites, while popular, attracted only 10 per cent of all countryside visits. These could be developed as 'gateways from which the public can explore . . . the wider countryside beyond, with all its features of villages, churches and pubs, as well as farmland and woods' (Countryside Commission 1987: 21). A subsequent report on *Visitors to the Countryside* (Countryside Commission 1991) stressed the uneven pattern of recreational use of the countryside. Although a few areas were under intense visitor pressure at certain times of the year, causing localised problems of congestion and erosion, the greater part of the countryside remained 'little visited beyond the use made by local people' (Countryside Commission 1991: 6).

It now seemed that there was considerable scope for rural areas to develop a visitor economy to compensate for the downturn in agriculture. Attractive landscape and rural heritage became valuable assets for place promotion, but a balanced, holistic approach was needed to ensure long-term growth and prosperity. In association with the Department of National Heritage (now Culture, Media and Sport), the Rural Development Commission and the English Tourist Board, the Countryside Commission (1995) established a set of principles for sustainable rural tourism. These principles were informed by critical discussion of a number of local schemes that demonstrated efforts to 'strike a balance between the needs of the visitor, the environment and local communities' (Countryside Commission 1995). The case studies included the Tarka Project, set up in a relatively remote and under-visited area of North Devon and based on places featured in Henry Williamson's cult novel about the life of an otter. The landscape varied from beaches, sand dunes and estuaries, the valleys of the rivers Tawe and Torridge and rose to high moorland at the edge of Dartmoor. Surveys showed, however, that 90 per cent of visitors to Devon stayed at or near the coast and only one per cent in the two valleys (Devon County Council 1988). Beef and dairy farming were in decline and the development of tourism was therefore encouraged. Sensitivity to the landscape and its wildlife was, however, paramount.

A primary motive of the Tarka Project was to demonstrate the value of the natural world as a resource for tourism, and hence encourage more resources to be spent on conservation measures such as improvement of water quality and habitat management. Visitors have therefore been encouraged to use sustainable transport, especially the Exeter–Barnstaple railway. The trail has also been accompanied by promotional leaflets, education packs, walks and cycle hire from stations. An extensive walking route and cycleway around the river estuaries was created and signposted as the Tarka Trail and local services and products were promoted through the Tarka Country Tourist Association. More recently, the support of visitors has been emphasised, with the Friends of Tarka set up to bolster local conservation. Reviewing its achievements, the Commission (1995: 70–3) concluded that the project had demonstrated the link between tourism and conservation, and had created a distinctive identity for the area. The trail had provided an important unifying theme, and surveys suggested that around 11 per cent of visitors had been attracted by media coverage. Nevertheless, the route had been designed primarily for long-distance walkers, and much of the initial promotion had been aimed at specialist markets such as the hiking press. With hindsight, more could have been done to reach 'mainstream' holidaymakers who prefer more casual recreation with a 'green theme'.

Over the past decade or so, public policy has put considerable stress on the enlightened common interest of the actors involved in local initiatives. Thus, partnerships between public, commercial and not-for-profit sector organisations have been supported to capture the benefits of growth markets in leisure and tourism which rural areas can absorb without detriment to the physical environment or host communities. By its very nature, voluntary collaboration depends on goodwill and the acceptance of common goals. In practice, much depends on the drive and persuasiveness of project managers, most of whom strive with limited resources to overcome conflicts of interest and develop a collaborative relationship between participants to develop programmes of action. The co-operation of key stakeholders, such as landowners and farmers, can never be assumed. Furthermore, the regulatory powers and resources of public agencies nurturing sustainable forms of leisure and tourism must be set in the broader context of market demand and the powerful commercial interests which finance and drive new development in rural areas.

The more confident and mobile middle classes dominate the Commission's category of 'frequent visitors': the upper quartile who 'tend to live either in or near the countryside, and own good quality houses and perhaps two cars' (Countryside Commission 1991: 6). For this group, the countryside can play an important part in leisure opportunity and the very quality of life. They also tend to represent the target market for a wide range of goods and services that reinforce the idealisation of rural life (Bunce 1994: 68–76). Examples include 'country kitchen' food, drink, fittings and decor; furniture and soft furnishings in traditional designs and materials; and 'country casual' clothing and other branded products. Affinity with the countryside ideal is also expressed in the purchase of high-value offerings which provide access to it: four-wheel drive jeeps, caravans and campervans, time-share property and second and retirement homes. Harrison (1991: 159) stresses the opportunities for the socially and personally mobile to experiment, 'casting on and off of identities'. She highlights the rising consumer spend and influence of the identity-conscious service class, who may choose to participate in more elite sports such as polo, archery, shooting and fly fishing, or visit theme parks, medieval fayres, pop festivals and open-air museums. The leisure culture associated with the countryside is thus creating a demand for a service infrastructure, facilities and events that cater for a diversity of interests.

Those who are socially and personally mobile thus have freedom to take up residence, short term or long term, in a rural area that suits their lifestyle. Fees (1996) considers this concept in a case study of the North Cotswold town of Chipping Camden, illustrating the concept of the 'resident tourist' for whom the locale is a leisure backdrop to a 'real' life lived primarily elsewhere. In the 1980s and 1990s, for the first time since the Industrial Revolution, there was a net migration into rural areas. The places that have received this population growth are well beyond the Green Belts that were originally designated to contain commuters and absorb the physical expansion of urban areas (Hall 1988). This counter-urbanisation may, to some extent, reflect new spatial patterns of employment, where rural areas have been the net beneficiaries. But it is also a people-led phenomenon that can only be explained by the appeal of the rural residential environment (Halfacree 1997: 74). Where planning permission allows, rural areas are being colonised by people with city-based occupations, as well as those who acquire second or retirement homes as an expression of the preference of higher earners to live and enjoy their leisure time in pleasant green surroundings near to home.

The 'democratic' tone of the new policies for the countryside implies participation by a broader section of the population. Many schemes initiated by local authorities, as well as other public and voluntary sector organisations over the past decade, have demonstrated a keen awareness of the barriers experienced by disadvantaged groups and the measures required to overcome these. This has led to an increasing desire to improve accessibility and access within the constraints of available resources. In 1999, the Countryside Agency was established by merging the Countryside Commission and sections of the Rural Development Commission. The Agency will play a critical role in carrying forward the government's proposals to secure better public access to the countryside, especially over mountain, moor, common, down and heath. Its three main aims signal a new emphasis in public policy: to conserve and enhance the natural beauty of the countryside, to promote social equity for the people who live there and to help everyone, wherever they live, to enjoy it without damaging its future (Wakeford 1999).

The commercial desire of some rural stakeholders to attract high-spending visitors as well as resident or semi-resident tourists may, however, do little to promote social equity. Like its urban counterpart, rural gentrification may displace less affluent residents, and alienate the lower quartile of the population: typically those 'living several miles from the countryside in poor housing, and dependent on public transport' (Countryside Commission 1991: 6). Examination of the low participation of particular social groups may reveal more sinister issues of exclusion, including racism. Agyeman and Spooner (1997) challenge the construction of images of rurality which sustain a mythological and nostalgic white heritage. People of colour, as well as other groups may be considered 'out of place,' and thus denied a sense of attachment to the countryside. This 'whiteness' of the British countryside was aptly illustrated by Pollard, a member of the Black Environment Network, who sought to disrupt dominant notions of a 'white' rural heritage by situating black people in prominent locations within her photographs of the 'English' landscape (Rose 1995: 111–17).

Overview

Two powerful discourses have influenced the way in which the British countryside is experienced by urban visitors and the way in which valued landscapes are viewed as a common heritage. First, the Romantic move-ment's affinity with Nature in the round stimulated a full-hearted appreciation of rural landscapes by those who had acquired the necessary cultural capital. Second, a belief in 'rational leisure' encouraged recreation in the countryside not only to develop a healthier physique but also to refresh and improve the mind and rejuvenate and inspire the spirit in beautiful surroundings. Both notions emphasise the sanctity of the coun-tryside, contrasted with the unwholesome features of urban life. The countryside must be treated with due consideration and visitors discour-aged from importing the vices, activities and unwelcome modes of urban behaviour. The landscape thus offers a challenge for the 'untutored townsman' to learn, gain insight and to develop the body, intellect and soul.

Enjoyment of the countryside and public policy to promote outdoor recreation has, nevertheless, been shaped through the complex interaction of a rich diversity of social groupings, cultures and subcultures. The coun-tryside came to be seen as a valuable resource offering the benefits of fresh air and opportunities for healthy exercise to the urban poor. Such causes were championed by articulate social reformers, as well as through the action of working-class people themselves. The legal right of access to countryside and protection of its natural beauty also became a rallying point for middle-class recreationalists whose activities were often organ-ised through clubs and societies. Legislative reform was, however, slow to emerge, encountering considerable opposition from the rural estab-lishment, and often bringing disappointment to its supporters. In the post-war era, resistance to 'urban interference' in rural affairs has been expressed periodically by well-connected landowning interests. This has been witnessed most recently in the form of the Countryside March of 1998 which was orchestrated to defy the newly elected Labour govern-ment's commitment to both the ramblers' Right to Roam in open country and legislation to ban foxhunting.

If the countryside ideal emerged and evolved along with industrial society (Bunce 1994), it seems likely that it will evolve still further in the post-industrial era. Over the last half century, there has been a marked decline in the pioneering working-class spirit which drove the 1949 legislation, perhaps reflecting more deep-seated social, economic and cultural change, and the decline of working class identity (Blunden and Curry 1989: 252–3). Visiting the countryside is more popular than ever before, but gone is the mass participation of close-knit industrial communities in activities such as hiking which was characteristic of the inter-war period. Growth in car ownership has facilitated a spectacular expansion of leisure and tourism in rural areas, but there is considerable polarisation in the level of participation. Those without private transport – especially social groups in urban areas who are less well off, less well informed and less confident – tend to visit rarely, if at all. In contrast, the countryside has proved highly attractive to those who enjoy a high degree of social and personal mobility.

In many parts of rural Britain there is now a strong economic argument for attracting urban visitors. Strategies are being formulated to satisfy the demand for recreation away from established, well-known destinations. As a result, policies of containment and visitor management have conceded to policies that encourage exploration of places further 'off the beaten track'. Such rural spaces remain a labyrinth of dark corners and mystery, a rural heterotopia that allows visitors the freedom to stumble across unexpected meanings and memories (Urry 1995: 228–9). An optimistic interpretation of current trends suggests that the countryside of the twenty-first century will be more inclusive and accessible to urban visitors, with a range of local initiatives creating new synergies between leisure and tourism, farming and conservation. A more pessimistic view is that rural landscapes will become increasingly commodified and 'valued' in the mercenary sense of the word. Socio-spatial segregation will become more pronounced than ever in a 'designer countryside', constructed to accommodate paying guests and discouraging those who could arguably benefit most from rural leisure.

5 Representing landscapes
Literary and artistic ways of seeing

Introduction

The term 'landscape' suggests a scene from nature that has been appropriated and framed by the agency of human perspective. As Chapter 3
explained, the etymology of the word itself tells us that there have long
been associations between landscape and commodification. The influence
of the European landscape medium in the last two centuries has been
such that it is difficult to separate the appreciation of actual countryside
regions from their artistic representations. The everyday language used in
appreciating natural environments reminds us of this link: we talk of
'picturesque' areas, of 'scenery' and of places being 'picture-postcard-
perfect'.

This chapter will explore the close link between valued landscapes and
artistic representation by examining theories that engage with the aesthetic
relationship between people and their environments. Ideas that humankind
has, since primitive times, been preconditioned to prize particular terrains
or that we experience a physiological reaction to certain environments are
discussed. More recent studies that attempt to uncover the socio-cultural,
rather than biological context for a landscape aesthetic are then explored.

If a landscape is a cultural construction, nature framed and set aside
for contemplation, either through pictures, literature or in actual environments, then it is clear that a strong association must exist between
artistic representation, landscape appreciation and that other form of
consumption: tourism. Artistic representations and tourism have always
enjoyed a symbiotic relationship: the Romantic period is often described
as the heyday of artistically inspired tourism as hordes of visitors thronged
mountains, lakes and waterfalls in search of scenes already described or
painted for them. The influence of such representations is just as strong
today with television and film imagery enticing visitors to locations.

The Highlands of Scotland is a region that clearly exemplifies the link
between artistic representation, aesthetic appreciation of landscape and
tourism development. Both physical and cultural geographies are significant in shaping the leisure and tourism landscapes of the Highlands where

absolute and symbolic space interact to create an imaginative country. This chapter seeks to explore the philosophical, literary and artistic roots of its imaginative reconstruction. The area's history, as a developing tourist resort, will be set in the socio-cultural context of the Picturesque and Romantic movements. The influence of novelists and poets who effectively created a language for articulating the Highlands as a valued and ideologically laden playground is also considered.

A landscape aesthetic

The disciplines of philosophy and art history have addressed themselves to the question of why we are so fond of particular natural scenes and of their artistic representations. Coming respectively from the human and the artistic perspective, they have attempted to develop a landscape aesthetic to explain the strong link between humankind and particular landscape formations. K. Clark suggested in 1949 that the link between landscape and our ideas of beauty is inherent, at least within an English context: 'almost every Englishman, if asked what he meant by "beauty", would begin to describe a landscape – perhaps a lake and mountain, perhaps a cottage garden, perhaps a wood with bluebells and silver birches, perhaps a little harbour with red sails and white washed cottages; but at all events, a landscape' (Clark 1991: 231).

K. Clark's perspective, that beauty is embodied in nature that is separated from us as human beings and depicted in landscape art, is now considered to be innocent and overly simplistic in its disregard for the ideological implications of landscape (J. W. T. Mitchell 1994: 6). However, the seeds of his ideas were sown 200 years earlier by the philosopher Edmund Burke, who, it is suggested, was certainly attempting to devise a landscape aesthetic within his developing theory of the Beautiful and the Sublime (Appleton 1996; Bunce 1994). This theory no doubt drew on the earlier writings of Joseph Addison which proposed that natural scenery embodied particular values. These were either the beautiful, which he describes as ordered and cultivated, or the grand and august in nature which display magnificence and are usually found in mountains, open plains or the sea (Parks 1964; Schama 1995).

In 1757, Burke's *A Philosophical Enquiry Into the Origin of Our Ideas of the Sublime and the Beautiful* (1998) sought, to explain our reactions to certain natural scenes in terms of our physiological response. Thus, the Sublime was that which affected us because it induced fear and pain, albeit a pleasurable pain and contained fear, and the Beautiful was that which suggested to us pleasure and society. In landscape terms, the Beautiful was conveyed in domesticated scenes but the much more influential Sublime was represented by the dizzy ravines and awe inspiring cliffs found in scenes painted by such artists as Salvator Rosa and found in real terms in the Alps and the highland areas of northern Britain (Cosgrove 1984). Burke was conducting his Philosophical Enquiry during the period when philosopher Jean-Jacques Rousseau first began to deliberate on the relationship between humankind and nature in his *Discours sur l'Origine de l'Inégalité* (1755). This work, and his later novel *La Nouvelle Héloïse* (1761), describe the healing powers of nature and the peace to be gained from leaving behind the corruption of society (cited in Grimsley 1973). These ideas were gaining widespread currency and landscape paintings and prints became a popular means by which the redemptive forces of nature might be consumed.

The idea of the Sublime, and its manifestation in artistic representation and tourism activity, was incorporated into the influential theory of the Picturesque which is usually attributed to the Reverend William Gilpin. This retired schoolteacher made a series of tours throughout England between 1768 and 1776 and produced lasting souvenirs of his journeys in the form of sketches and landscape descriptions which were later published. These publications directed the reading public to those places worthy of visiting by using very specific aesthetic criteria. Appleton suggests that Gilpin may have been influential in directing public taste or may simply have been in tune with that taste; either way, his contribution is important in that he presented accounts of actual places and linked their physical characteristics to the emotional responses generated by these (Appleton 1996: 29). Gilpin says of the area around Derwent Water, 'Nothing conveys an idea of beauty more strongly, than the lake; nor of horrour than the mountains; the former lying in the lap of the latter, expresses in a strong manner the mode of their combination' (Gilpin 1786 quoted in Appleton 1996: 29).

However, despite the immediacy of this response, the Picturesque also seemed to distance the viewer, spurred on to visit the sites depicted in Gilpin's volumes, by encouraging them to see the landscape as if it were actually a picture. The philosophy of the Picturesque was one of framing the landscape, sometimes literally and often within the Claude glass, as described in Chapter 3. This 'framing' heightened the experience and concentrated the viewer's gaze (Bell 1993: 12). Punter explains the difference between this and the Sublime in the importance of the boundary: 'The Picturesque frames roughness and variety; the Sublime has to do with being overwhelmed, surprised, being taken out of one's frame by a scene' (Punter 1994: 223). As a way of seeing, the Picturesque directed the popular gaze towards visualising nature as a series of framed scenes which prefigured the tourist way of seeing. As Chapter 3 has already outlined, these emerging views of leisure and tourism landscapes were made possible by ever increasing improvements in transportation but also, and importantly, gave us a vocabulary for explaining the visual appeal of landscape (Urry 1995: 196).

Later, the specificities of the Sublime and the Picturesque became subsumed in the political, artistic and philosophical revolution of the Romantic Movement which was to have an enduring impact on the way that landscape was perceived. The Romantic Movement heralded a shift in artistic and philosophical values away from the rational and towards an understanding of the supremacy of nature as a guiding force for humankind. Rousseau's influence has already been stated: as a leading exponent of Romantic philosophy, he believed that humans could only find their true identity through nature and that a return to the natural state would ensure happiness and fulfilment (Grimsley 1973). The philosophy of Romanticism gained momentum with the French Revolution and its elevation of the common man and woman through freedom: it was also fed by a reaction against the continuing influences of industrialisation and the values of science and economics that such a system promulgated.

Artistically, the Romantic Movement was conveyed to a growing middle-class audience through the works of the English Romantic poets, Blake and Wordsworth being the most prominent. Through their work, the reading public were encouraged to see nature as a fit subject for poetry and their poems offered healing and spiritualism through the description of natural scenes and, significantly, the emotions conjured up by those scenes. Humans were seen as being at their most natural when alone in the wilderness, as opposed to the artificial society created by the demands of the industrial system which distracted us from our true purpose and path to fulfilment: 'the moral order of society has been replaced by an economic order. If morality is to be discovered it must be in nature ... Thus romanticism seeks divinity in the mountains and sermons in stones' (Cosgrove 1984: 232).

The philosophical underpinnings of Romanticism were soon engulfed in the wholesale acceptance of visual Romantic values by the middle-class Victorian consumer of landscape paintings, engravings, watercolours, poetry and tourism, and the fashion for the appreciation and development of 'taste' in Romantic scenery became well established (Urry 1995). Visits to the Alps, the Lake District and the Highlands of Scotland originally encouraged through the Picturesque sensibility, became even more popular for consumers sensitised to the Romantic through visual consumption: 'by Victorian times, the romanticisation of nature had reached heights fuelled as it had been for several decades by a rising tide of poetry, art and music which drew upon nature for its creative inspiration' (Bunce 1994: 26). The Romantic ideal was translated into a popular landscape taste which is still discernible today.

Thus, the pure Romantic values of the immediacy of the human response to landscape became merely a fashionable taste, fanned by the socio-cultural and technological context within which it flourished. And despite the work of earlier philosophers and critics in trying to establish 'the nature and origins of the affective bonds between the conscious subject and its visible natural surroundings' (Cosgrove 1984: 17), more recent commentators have explained our aesthetic response to landscapes and their representation and consumption in sociological, rather than more atavistic terms. Our landscape ideals reflect the wider context of our existence that determines our reactions to particular landscapes and our consumption of them as readers, viewers and tourists.

The socio-cultural context

The argument that our affinity for landscape is socially, culturally and historically contextualised has been well rehearsed in recent years by commentators who closely link the creation of a landscape aesthetic to the forces of industrialisation or earlier changes in land use and ownership (Bourassa 1991; Bunce 1994; Cosgrove 1984; Schama 1995; Urry 1995). Cosgrove and Bourassa agree that, once land ceased to become the site of everyday labour for the majority, it became a commodity and subject to manipulation by artists: 'as the intimate tie between land and its users was severed with the development of capitalism, the idea of landscape arose. In other words, it became possible to distance oneself from the land so that it could be viewed as a landscape' (Bourassa 1991: 3–4). Bunce describes this process of land-to-landscape as the creation of a 'countryside ideal', the inevitable outcome of industrialisation, further strengthened by widespread consumption of 'armchair countrysides' by the growing middle classes (Bunce 1994).

Cosgrove argues for the social construction of landscape and sees it as a particular class-based way of viewing the world, rather than a mimetic reflection of it:

> Landscape, I shall argue, is an ideological concept. It represents a way in which certain classes of people have signified themselves and their world through their imagined relationship with nature, and through which they have underlined and communicated their own social role and that of others with respect to external nature.
>
> (Cosgrove 1984: 15)

On a more individualised basis, Schama takes the view in his exploration of *Landscape and Memory* (1995) that the power of memory itself is important in shaping our response to landscapes. This memory is of course highly personalised for each individual but he also accepts that these memories may have a collective base and come from the 'inherited tradition' reaching back to images of Eden as a communal source of imagery (Schama 1995: 12). This notion of collective memory has already been discussed in relation to landscapes giving rise to the monument and the spectacle (Chapter 2) and is discussed further in relation to both heritage and urban landscapes in Chapters 6 and 8 respectively.

Appleton's work *The Experience of Landscape* (1996) also examines representation from the perspective of its social construction, but he utilises behavioural theory to explain the nature of the 'inherited tradition'. This more atavistic explanation of our affinity for particular landscapes, he calls 'prospect/refuge theory'. This theory links our basic human needs for shelter and a vantage point to our tastes and preferences in landscapes and landscape representations. Thus an 'ideal' landscape would offer the shelter of a shady knoll and the prospect of a hilltop, suggesting far horizons. The behavioural basis of Appleton's work owes some of its inspiration to Burke and his early explorations into our immediate responses to the beautiful and the sublime. In both theoretical frameworks, people are seen to have preconditioned and physical responses to landscape but in Appleton's case, the physical has become deeply embedded in the cultural.

Landscape and imaginative reconstruction

What the various landscape approaches described above have in common is the understanding that, far from being a 'natural' relationship, the affinity that humans have for landscape is predominantly a (by-)product of the imagination, shaped by a variety of social and cultural constructs. Indeed, the notion of 'nature' itself is a human creation which has undergone various shifts in meaning through the development of humankind (Gold 1984). Urry suggests that people have interacted with this construct called 'nature' in various ways throughout history, from the stewardship of land, to exploitation, to scientisation, culminating in visual consumption: the contemporary mode of interaction (Urry 1995: 174). Urry's brief 'history' of our relationship with nature is, of course, Eurocentric: there are many cultures today who demonstrate an interaction with their environment that does not follow this unilinear development. However, the dominant mode, from a Western perspective, is that of consumption. Thus, landscapes themselves are not 'natural' but indeed highly artificial constructions which have been created for particular types of use, predominantly tourism. The link, therefore, between artistic mediation and the touristic consumption of landscape, as noted earlier, is clear.

Furthermore, it has been suggested that we learn how to appreciate nature (and thus we learn our tourism preferences) through art itself. Bourassa quotes Apollinaire, writing in 1949 on this topic: 'without poets, without artists . . . the order which we find in nature, and which is only an effect of art, would at once vanish' (quoted in Bourassa 1991: 11). Thus, we learn to see particular forms of beauty in the countryside because we are accustomed to seeing them in galleries, extolled in poetic descriptions and, later, through the media of print, advertising, television and film. Gombrich (1966) echoes Apollinaire in his statement:

> I believe that the idea of natural beauty as an inspiration of art . . . is, to say the least, a very dangerous oversimplification. Perhaps it

even reverses the actual process by which man discovers the beauty of nature. We call a scenery 'picturesque' ... if it reminds us of paintings we have seen. And to the painter, in turn, nothing can become a 'motif' except what he can assimilate into the vocabulary he has already learned ... If Patinier, for instance, really embodied reminiscences of the Dinant scenery in his paintings, if Pieter Brueghel really found the Alpine peaks inspiring, it was because the tradition of their art had provided them with a ready visual symbol for steep isolated rocks which made it possible for them to single out and appreciate these forms in nature.

(quoted in Appleton 1996: 208)

Talking more specifically about the literary heritage of landscape representation, Pocock suggests that:

Our cognitive frame of reference for viewing reality, therefore, may well be to varying degrees a literary frame of reference. The potency of literature as a creative force in this sense is seen in the way many parts of Britain are approached, 'seen' and remembered through the eyes of poets and novelists, whether we be native or visitor.

(Pocock 1981: 13)

This process of imaginative reconstruction of places is amply illustrated in the case of the Lake District whose development as a tourist resort was a direct consequence of its 'discovery' by travellers, sketchers, water colourists and poets. These early visitors were happy to disseminate images of the Lakes to an appreciative and Romantically inclined public. Urry has suggested that it was the artists and writers themselves who created the Lake District out of an unknown part of Northern Britain and allowed it to be seen through the tourist gaze, whether through the actual framing of the Claude glass or through the literary or artistic framing of the viewer's earlier consumption of landscape representations (Urry 1995). The appeal of Romantic ideology, the increased mobility of a growing middle-class audience and the development of a fashionable taste for upland areas, first popularised by the upper classes' affinity with the Alps as part of the Grand Tour, created a tourism district that was first experienced and consumed at home as an 'armchair countryside' (Bunce 1994) and then experienced in the raw as a destination, already imaginatively reconstructed in the eye of the beholder.

In his text *The Beaten Track: European Tourism, Literature and the Ways to Culture 1800–1918*, Buzard suggests that there has always been a strong link between tourism and artistic representations, especially literature: 'tourism and writing have been reciprocally related. In this work I attend ... to tourism's resourceful appropriation of literary texts and reputations to enhance its excitement and give shape to its new cultural aspirations' (Buzard 1993: 13). Furthermore, it has been argued that it is in fact the literary and artistic associations which give a place meaning (Herbert 1995; Pocock 1981). Grand Tourists and early cultural tourists were reported as taking with them on their travels, 'a Murray for information and a Byron for sentiment' (Buzard 1993: 120), thus enabling them to find places associated with the poet using John Murray's famous guide book, and then experience Byron's works in situ.

Artistic and literary tourism consequently has a long history. Andrews' work follows the *Picturesque Tours of England and Scotland* which led visitors on a pre-ordained route through places of literary and artistic association in the 1700s (Andrews 1989). In the 1890s there were coach tours to Constable Country in the Stour Valley in Suffolk (Herbert 1996) and Ousby (1990) has described at length the tourism industry that quickly emerged in Shakespeare's Stratford-upon-Avon. Alfred, Lord Tennyson is supposed to have been far more interested in Lyme Regis as the fictional setting of Jane Austen's *Persuasion* than in its more historical connections – a predilection towards literary tourism perhaps fitting in a Poet Laureate (Ousby 1990: 21).

This tradition of imaginative reconstruction has continued today, although the literary and artistic tourist has been replaced, in number at least, by the 'teletourist' seeking places already made famous and accessible to them as locations for television dramas and films (Gold and Gold 1995; Riley 1994; Schofield 1996; Tooke and Baker 1996). The British Tourist Authority has recently provided new maps of Britain for the cultural tourist keen to seek out locations. 'The Movie Map' and 'Literary Britain' point out places associated with film, television, writers and fictional characters and provide an alternative cultural geography of the British Isles.

The Highlands of Scotland

> The Highlands, then, are imaginary
> (Womack 1989: 166)

The case of the Scottish Highlands illustrates the close relationship between cultural representations of landscape and the creation of a resort for touristic consumption. The word 'myth' has been used to describe certain Highland/Scottish national and landscape ideals and values, and this word conveys much of the constructed and artificial nature of this land of the imagination (Hills 1994; Pringle 1988; Withers 1992; Womack 1989).

Chapter 6 goes on to illustrate ways in which, as a tourist destination, Scotland has been endowed with an extremely rich iconography that is recognisable across the globe (Butler 1998; Hills 1994; McCrone, Morris and Kiely 1995). These well-known icons include tartan, heather-clad hills, mist-strewn lochs, bagpipes, whisky, stags and a stalwart and colourful people and are often seen as a shorthand for the nation of Scotland as a whole. Landscape images are an important part of the iconography and are replete with myth and meaning. Valued landscapes and their representations are an important means by which a nation creates a sense of its collective identity: 'every mature nation has its symbolic landscapes. They are part of the iconography of nationhood, part of the shared ideas and memories and feelings which bind a people together' (D. W. Meinig quoted in Osborne 1988: 162). The Scottish Tourist Board uses 'land-scape' to describe the main 'product' that is Scotland and which can be readily recognised outside the country (Cano and Prentice 1998). However, in the case of Scotland, the elements which together form this symbolic landscape are all taken from a very specific region; the Highlands. This region has come to stand for the whole country, as have the Highland people come to represent all Scots, whether in tourism promotion or in wider ideas about nationhood (Gold and Gold 1995; McCrone, Morris and Kiely 1995; Womack 1989). The cultural centrality and mythical nature of this representation is undoubtedly the result of a complex series of political and social circumstances. As the following case study will demonstrate, however, one of the most significant factors has been the picturing of Highland Scotland. This literary and artistic representation created an enduring image of the area which, in turn, has shaped its socio-geographic and cultural structures (Hughes 1992).

As Chapters 3 and 4 have indicated, literary and cultural tourism has been a major feature of travel since people began touring for pleasure in any significant numbers. However, like the Lake District, the Scottish Highlands have been the subject of a large and influential series of representations that still impact on the way the nation is seen and promoted. The Highlands were not perceived as a distinct geographical area until after the 1300s and their separation from the Lowlands is a cultural rather than a physical geographical phenomenon now usually associated with the Gaelic-speaking areas or Gaedhealtachd (Withers 1992: 144). In terms of being known to the rest of the world, Scotland was seen, well into the 1700s, as a barbaric and hostile territory with uncivilised tribes making raids upon the south. After the Act of Union in 1707, and the quelling of the clans at Culloden in 1746, however, the Highlands in particular seemed to undergo a transformation which rendered them open to a range of cultural representations: in short they became a mine of iconography from which much could be quarried.

The cultural transformation of the Highlands in the popular imagination was the result of a number of factors, one being the Napoleonic wars of 1796–1815 which effectively closed off the continent to potential Grand Tourists thus forcing them to turn their sights closer to home. A wider appreciation of the landscapes of the British Isles, now that much of them had been tamed, was also a contributory factor. The Picturesque way of seeing, as instructed by the Reverend Gilpin had by then firmly taken

hold and the particular geomorphology of the Highlands, with their mountains, lochs, glens, waterfalls and brooding climate tuned in well with this
(Andrews 1989).

Prior to this transformation, when Dr Johnson and Boswell reported
on the uncouth natives and barren landscapes during their famous and
ambitious tour in 1775, these particular features were not deemed to be
worthy of viewing or sketching. However, by the later processes of cultural
representation and dissemination, consumers 'learned' to appreciate these
landscapes. In his introduction to *Improvement and Romance: Constructing the Myth of the Highlands* (1989), Womack describes the reactions
of an English traveller, Edward Burt, to the heather-clad mountains
near Inverness in 1730. The Englishman thought them dismal and
gloomy because, as Womack suggests, 'Burt doesn't know what heather
"means" for him, the plant is innocent of romance. This is because when
he was looking at it, the romance had not yet been invented' (Womack
1989: 1).

Similarly, the observations of Dr Johnson on the barrenness and bleak
aspect of the country contrasts well with later and more ebullient descriptions. But why did the Picturesque and late Romantic sensibilities take
such a hold and appropriate so wholly, the Scottish Highlands? The answer
lies within the pictorial and literary representations of the nation which
began to flow from pens and brushes from the late 1700s onwards.
Enlightenment thought about the progress of human society laid special
emphasis on the notion of the 'noble savage' and the Highland population, once feared and unknown, now came to represent a particular stage
in human development – in effect, a shred of living heritage still acting
out ancient rituals in a timeless landscape. Rousseau's ideas on the
simplicity and natural happiness of 'primitive man', in contrast with
'civilised man', also lent weight to this celebration of the Highlander. With
this elevation in place, and set against the rapid technological changes of
an industrialising nation, it is perhaps easy to understand why the cult
of Ossian, one of the most important literary influences on Highland
tourism, took such hold.

Ossianic tourism

The story of this famous literary hoax has been told in several places and
the link between the enthusiastic response to these poetic 'fragments' from
an allegedly ancient Bard and the increase in Highland tourism has been
demonstrated (Andrews 1989; Durie 1994; Gold and Gold 1995; Womack
1989). A Scottish schoolteacher, James Macpherson, is credited with
creating these translated Gaelic poems which were attributed to the blind,
harp-playing Bard, Ossian, in the third century AD. Macpherson published
*Fragments of Ancient Poetry Collected in the Highlands of Scotland and
Translated from the Gaelic or Erse Language* in 1760 and later works,

including *The Works of Ossian, Son of Fingal* went through twenty-five editions in thirty years (Andrews 1989: 203).

The sceptical Dr Johnson was not convinced of the authenticity of the *Fragments* although they were supported by a range of influential enthusiasts including Hugh Blair, Professor of Rhetoric at the University of Edinburgh, who wrote a Critical Dissertation on the subject, Mendelssohn, who wrote *Fingal's Cave* in the poet's honour, and Napoleon, who purported to carry the poems into battle with him (Scott 1994: 365). Whether they were believed to be real or not, the works were enthusiastically read and later recited by tourists alongside appropriately romantic waterfalls and on mountain peaks. Put into a literary context, the *Fragments* became popular because they provided a break with the measured epigrams of the period and reflected a landscape and ancient traditions full of chivalry and primitive nobility: 'both in cadence and imagery, these "songs of other times" distilled the influence of the landscape and the climate, and their primitive idiom was a thrilling contrast to the polished social poetry of the Augustans and their successors' (Andrews 1989: 202).

As tourists on the trail of the Ossian myth began arriving in the Highlands in some numbers to view the landscapes epitomised in the *Fragments*, specific attractions were created to satisfy the desire to see actual (or potential) sites associated with the Bard. The poetry itself was rather vague on geographical location which only added to the appeal for tourists to decode the work and find their own landscape associations (Womack 1989: 109). Just as Chapter 6 makes reference to Stirling as forming a 'gateway' to the Highlands, so Dunkeld was previously described in this way. The Dunkeld gateway was generally agreed to have some clear associations with Ossian; the Narrow Glen for example was thought to be the location of his grave. Important sites linked to the poet's life and death show a distinct leaning towards brooding environments as he was supposed to have been born in the bleak and magnificent surroundings of Glen Coe.

The link with Dunkeld prompted the Duke of Atholl to create on his estate 'what could be described as the equivalent of a modern theme park' dedicated to Ossian (Gold and Gold 1995: 56). The Falls on the River Bran at Dunkeld had already become a picturesque tourist attraction in their own right, the Reverend Gilpin having been particularly impressed by the sublime grandeur of their forty foot drop (Andrews 1989: 214). The Hermitage, or summer house overlooking the Falls was remodelled in 1783 as Ossian's Hall, with a large portrait of the Bard in the entrance vestibule which swung back to reveal a windowed room. This room, which looked directly onto the Falls, was mirrored on ceiling and wall to reflect the churning of the water. The roar of the Falls would also have added greatly to the visitor experience. Further up the cascade the less formal structure of Ossian's Cave was also created with one of the Bard's poems etched onto the walls. These literary additions to an already picturesque scene were greeted with enthusiasm by their many visitors, although more cynical commentators such as Dorothy and William Wordsworth, touring in the Highlands in 1803, were amused by the Duke's 'distressingly puerile' attempts to create a sublime visitor attraction (Andrews 1989: 216).

Referring to the work of Leneman (1987), Gold and Gold (1995) suggest that the creation of an Ossianic literary landscape in the Highlands underlines three distinct processes: an imaginative reconstruction of landscape, of the people supposedly inhabiting it, and the creation of a literary frame of reference for tourism:

> Ossianic poetry had a threefold effect. First it gave people a new way of looking at sublime, wild and desolate scenery. Secondly, it encouraged them to view landscapes in association with Ossian, peopling the landscape with their imagination. Finally, it supplied a new way of seeing the highlander. If the landscape had not changed over the centuries, then the highlander must be the same noble creature as of yore.
>
> (Gold and Gold 1995: 54)

This attitude towards the Highlander as an integral part of the romanticised landscape is seen as one of the ironies of Scotland's cultural transformation (McCrone, Morris and Kiely 1995). A race once reviled and feared now came to represent the truth and nobility of our ancestors – in short, as Womack suggests – the Highlander becomes equally transformed and 'no longer as different from ourselves but as what we once were' (Womack 1989: 23). Of course at the time of this transformation, the power of the clans had been dissolved and the sturdy Highlander was soon to become a rare figure in reality, although a popular figure in romantic representations.

The Highlander in the picture

As part of the forced dissolution of the feudal system of the Highlands, traditional clan chiefs were replaced by absentee landlords who instigated the Highland Clearances of the 1800s. The population of the Highland glens was forcibly cleared out and relocated to the coastal areas of Scotland or to the colonies of New Zealand, Australia and Canada in order to make room for more profitable sheep farming (Butler 1998). Thus, much of the Highland population was forced to turn to fishing as a means of supporting themselves in their new communities in the less fertile coastal areas. The people of the Highlands, once considered irreligious and lawless, had been rendered impotent by their defeat at Culloden in 1746 and could now been seen as an anthropological curiosity, or as 'a primitive people with the charm of an endangered species' (Trevor-Roper 1983: 25). The ruggedness and grandeur of the Highland landscapes was imbued in the psyche and traditional lifestyle of the people who became an objectified part of that landscape. The empty Highland glen became a romantic ideal in artistic representations of the area, and yet the supposed 'naturalism' of this representation hides the story of a forced relocation (McCrone, Morris and Kiely 1995: 59; Pringle 1988: 153).

The appreciation of Highland imagery and values as representative of an entire nation was another important stage in the transformation of the Highlands. The social elite of the Lowland population adopted the myths and symbols of the Highlands as evidence of a Scottish culture distinct from that of England. As Trevor-Roper (1983) has demonstrated, the most potent symbol of the Highlands was the distinctive Highland dress. The plaid, a length of woven material folded and belted around the body and which later became the phileabeag or 'little kilt', was originally a cheap and useful garment for clansmen accustomed to scrambling over hills and sleeping out in the glens overnight. As a distinctive symbol of the clans, this traditional Highland dress was prohibited in 1746 after Culloden and the prohibition was only repealed in 1780. The rarefied version of this plaid, produced in brightly dyed tartan designs, became a status symbol for the Lowland elite who would often have their portraits painted depicting them in this colourful and exotic garb:

> Anglicised Scottish peers, improving gentry, well-educated Edinburgh lawyers and prudent merchants of Aberdeen – men who would never have to skip over rocks and bogs or lie all night in the hills – would exhibit themselves publicly . . . in a costly and fanciful version of that recent innovation, the philibeg or small kilt.
>
> (Trevor-Roper 1983: 24–5)

The final stamp of approval of the once reviled Highland dress was the wearing of the kilt by George IV on his visit to Edinburgh in 1822, the first visit to Scotland by a reigning monarch since 1633. The pageantry of this event was stage-managed by Walter Scott, who did much to further popularise Highland mythology in his novels and poetic works. The direct link between the wearing of tartan and the Highland population had now been made and was perpetrated through literature and paintings of an area where a kilted Highlander would be an accepted part of the land-scape's iconography despite the fact that the male Highland peasantry at least had long taken to wearing trews as a more practical form of dress (Trevor-Roper 1983: 24). The dramatic and romantic plaid was however used to denote all things Scottish: Womack (1989) notes that the stage tradition for dressing Macbeth in plaid began in 1773, although this garb would have been unheard-of in a medieval Scotland (Womack 1989: 144). John Martin's painting *Macbeth* (1789–1854), in the collection of the

National Gallery of Scotland, shows 'the Scottish play' receiving a similarly dramatic setting and costume.

One of the most well-known evocations of the Highlander in the landscape is William Wordsworth's famous poem 'The Solitary Reaper' which was published in 1805, following his Scottish tour with his sister Dorothy in 1803, although the original inspiration for the poem came from his friend Thomas Wilkinson's travel journal (Andrews 1989: 228). The poem describes the poet coming upon a solitary, noble Highland lass at work in the fields and the various elements of Romantic appreciation of the Highland landscape seem to be made flesh in the figure of the Reaper. First, she is described as being part of the landscape, filling the 'Vale' with her song and becoming assimilated with it. Second, she is differentiated from the poet and the reading public by speaking in a tongue not readily understood, and thus she is linked with images of exoticism and distant travel. Finally, the reaper is strongly associated with the past as Wordsworth puzzles over the subject of her song, wondering if she sings of melancholy past times in her plaintive Gaelic. In 'The Solitary Reaper', the poet puts himself firmly in the role of the tourist and the poem is as much about his response to the Romantic landscape of the Highlands as it is about the Reaper herself who is both a deliberate and Romantic enigma. Womack suggests that the poem is 'the representative lyric of Wordsworth's highland tour ... His characteristic activity as a tourist is pausing to decipher a wayside monument' (Womack 1989: 114). Thus, the Highlander is both a symbol of, and monument to, a past way of life.

The Highlands of Walter Scott

The 'solitary Highland lass' became a potent symbol and well-known figure for tourists to the Highlands but her fame was soon overtaken by another literary heroine, around whom a mini tourism industry was formed: Sir Walter Scott's *The Lady of the Lake* (1810). The role that Walter Scott has played in creating an enduring image of Scotland far beyond Scottish shores has been described in various texts (Butler 1998; Gold and Gold 1995; Seaton 1998; Womack 1989). His centrality in the creation of Scotland's unique iconography is not questioned. Gold and Gold (1995), for example, suggest that he was a 'one-man Scottish Tourist Board' (quoted in Seaton 1998: 17) although it has been suggested that Scott did not create Scottish tourism but rather used the Romanticism that his works fed upon (McCrone, Morris and Kiely 1995: 60). Either way, his novels and long poems, which he wrote between 1814 and 1832, recreated a world of historical detail and chivalrous romance that appealed to many readers, so much so that his works were translated into most European languages, became popular stage productions and sold millions of copies throughout the nineteenth century (Gold and Gold 1995: 65).

What is important about Scott's work in relation to the imaginative creation of the Highlands as a tourist resort is his use of landscape, peopled with romantic characters. Unlike the works of Ossian/Macpherson, Scott's works were geographically accurate and utilised settings that he knew well. As Watson suggests, Scott and Wordsworth, as Romantics, both have a very strong sense of landscape and feeling, rather than simply viewing:

> Both celebrate it as something to be felt, Wordsworth more profoundly as a source of happiness and moral power, Scott as connected with romantic events and his nation's history. Peopled with heroic actions, the landscape became for Scott a source of power and pride.
>
> (Watson 1970: 127)

His sense of geography was perhaps more acute than his historical accuracy, and he used the legends and myths he collected on his travels throughout Scotland as a source of historical detail. One of his literary devices (used, for example, in *Waverley* and *Redgauntlet*) was the incomer, usually English, to whom the history and topography of Scotland had to be explained. This allowed the reader and potential tourist to identify with the main character and gave Scott free rein for explanation and detail that he often gave in the form of very full footnotes which lent an air of authenticity to the works.

As mentioned earlier, *The Lady of the Lake*, Scott's long poem written in 1810, is often credited with popularising the Trossachs in the western Highlands as a tourist destination (Seaton 1998: 17; Watson 1970: 126). In fact Womack suggests that 'the poem was received as a literal invitation to visit its locations. It made the Trossachs into the prime Highland tourist site overnight' (Womack 1989: 155). However, Loch Katrine, the main location of the poem, had become a popular and romantic destination before the publication of *The Lady of the Lake* (Gold and Gold 1995: 69): the poem's role was to transform the area into a literary construction. Visitors arrived armed with copies of the poem to seek out locations and read passages in appropriately romantic settings in much the same way as Ossianic tourists had read and appropriated selected parts of the Highlands. Soon the trappings of tourism arrived at Loch Katrine: guided tours in four-horse wagons would arrive daily at the site equipped with drivers specially trained to recite passages from the poem. Indeed, American tourists were so numerous that their flag was flown at the local hotel (Gold and Gold 1995: 76). So great had been Scott's power over this part of the Highlands that landowners tried to persuade him to set his work in their districts to create a similarly lucrative tourist boom

(Durie 1994: 495). A clear indication of the strong connection between imaginative geography and actual topography is demonstrated in the fact that 'Ellen's Isle' from the poem is still marked today on Ordnance Survey maps (Womack 1989: 156).

The landscapes that Scott created in his works have been likened in their light and movement to those of J. M. W. Turner (Watson 1970) who was commissioned to create the illustrations for a two-volume book on *Provincial Antiquities: Picturesque Scenery of Scotland* (published 1819 and 1826) which was commercially unsuccessful (Butler 1998: 127). The link between Scott's work and other art forms is also demonstrated in the ease in which his works were turned into operas. Donizetti's *Lucia di Lammermoor*, Rossini's *La Donna del Lago* and Verdi's *Aroldo* all utilised the stirring events and colourful backdrops of Scott's work to dramatic effect (Gold and Gold 1995: 75).

Tourists also wanted to visit places associated with Walter Scott's life. Abbotsford, his fanciful baronial seat in the Borders, became a visitor attraction in its own right and remains so today (Durie 1994). Thus, Scotland's literary tourism exploited sites both actual and imaginary and both were powerful magnets for visitors.

Travelling in the Highlands

Tourists had started to come to the Highlands in ever increasing numbers, prompted by the pens of Scott, Macpherson, Wordsworth and other travellers and writers. Before the 1700s, travel in Scotland had been expensive, dangerous and difficult for the English market who were later to become the main source of visitors. Macleod noted in 1934 that:

> In the 17th and 18th centuries, the Islands and Western Highlands were more vaguely known to the people of London than Patagonia or Alaska is today to children in remotest Hebrides. A journey to the Scottish Western Isles was then looked upon as one of the most formidable feats of travel which an explorer could undertake.
>
> (quoted in Seaton 1998: 8)

However, thanks to the popularisation of Scotland through its imaginative reconstruction, by the 1820s *The Inverness Courier* was reporting on the large numbers of tourists entering the Highlands and 30,000 visitors were supposed to have been generated specifically by the publication of *The Lady of the Lake* in 1810 (Seaton 1998: 8). Travel to the Highlands had, of course, become easier for the tourist after the development of the steamer services on the west coast in the 1800s and by the advent of the railways: by the 1890s there were three railway companies providing routes between London and Scotland (Butler 1985).

The tourism potential of these services was fully realised by Thomas Cook, who had been arranging excursions into the Highlands as early as 1846 with groups of 350 travellers (Durie 1994: 496). This ambitious early foray, using a complicated array of specially arranged trains and steamers, was something of a disaster for Cook but he learned from these initial mistakes and two-week-long tours to Scotland were soon to form the basis of his business (Buzard 1993: 52–3). By 1861, Cook claimed to have personally escorted 40,000 visitors on his 'Tartan Tours' to Scotland, many of whom visited Scott's locations. Cook's promotional material drew heavily on literary allusions and it was expected that his excursionists would be familiar with Scott's work (Gold and Gold 1995: 103). Cook's tours to Scotland, with their attention to detail, meticulous travel arrangements and guidebooks, did much to bring the Highlands to a wider audience than the earlier elite cultural tourist and, in particular, allowed women to join the tours in safety (Butler 1985).

Royal patrons

A final important strand in the imaginative reconstruction of the Highlands was the very specific image created and disseminated by one extremely influential Highland tourist – Queen Victoria. Much has been written about 'Balmorality' and Victoria and Albert's Highland idyll (McCrone, Morris and Kiely 1995; Pringle 1988). Certainly, the baronial-style castle they created on their Highland estate in Deeside in 1852, furnished as it was in Balmoral tartan and stags' heads, did much to strengthen the Highland myth. This myth was exaggerated further by Victoria's *Highland Journals* and the specially commissioned portraits of their Highland life. For example, Sir Edwin Landseer's commissioned paintings of the royal couple at Deeside are replete with Highland iconography – *Queen Victoria sketching at Loch Laggan* (1847) and *Queen Victoria meeting the Prince Consort on his return from deer stalking* (1850) include romanticised landscapes, stags, lochs and deferential kilted ghillies as well as the royal family themselves in suitably Highland garb (Pringle 1988: 143).

Victoria and Albert themselves were well-versed in the Highlands of Sir Walter Scott: under Scott's influence they learned to appreciate the rugged landscape and to further popularise 'Royal Deeside' as a fashionable tourist resort. These royal connections are still used to promote this part of the Highlands as the current *Holiday Guide to Grampian Highlands, Aberdeen and the North East Coast 1999* demonstrates with its slogan of 'Royal Deeside, Whisky, Castles and Coast.'

A literary way of seeing

Thus, it can be argued that a literary way of seeing has created the Scottish Highlands that still exist strongly in the imagination and that visitors seeking the myths and romance of the Highlands are doing so through the eyes of Scott and Macpherson and associated followers, even if they are unfamiliar with the texts themselves (Scott 1994: 368; Gold and Gold 1995: 83). Scotland has therefore been subjected to an aesthetic framing that Craig (1982) suggests is specifically literary. Literary representations of Scotland have had particular potency in the collective imagination because:

> conditions in nineteenth-century Scotland would not let works of literature rest within the realms of literature. The worlds described by Scott . . . became the foundation of myths of national identity in a country whose individual identity had been swamped by its incorporation into the United Kingdom.
>
> (Craig 1982: 9)

The internal and external view of the nation has been created by what R. W. Butler describes as 'a Who's Who of English Literature – Defoe, Johnson and Boswell, Burns, Dickens, Coleridge, Wordsworth, Southey, Tennyson and Scott' (Butler 1998: 124). Their travellers' tales, poetry and fiction became a reality for those who visited Scotland. Thus, Highland tourism has, to a certain extent, always been literary tourism (Durie 1994: 495).

The tourism product of Scotland has, of course, become more diversified since its development in the eighteenth and nineteenth centuries. Sports, including hunting, shooting, fishing, walking and skiing, have become popular holiday activities and cities and towns now market their own particular urban attractions, for example Edinburgh, the 'Festival City' and Glasgow as 'Cultural Capital'. Yet the Romantic myths have remained the strongest pull in terms of tourism promotion and the way in which the nation, and particularly the Highlands, are seen (Gold and Gold 1995: 112; Seaton 1998: 15). Writing in 1934 of the quest for the true *Heart of Scotland*, G. Blake suggested that the motivation of the tourist in Scotland had hardly changed in a hundred years:

> the average traveller suffers under the illusions of the literary legend established during the nineteenth century. He is a man who is, most naturally, concerned to see what delights the eye and stirs the imagination . . . and we find him in the largest numbers in the Highlands . . . looking anxiously for every cave and corrie that has possible associations with Bonnie Prince Charlie and Alan Breck and heroes dead so long ago.
>
> (Blake 1951: 5)

Indeed, as R. W. Butler notes, the itinerary of places that tourists visit in the Highlands has remained largely unchanged over the centuries as only about 50 miles of new roads have been built in the region in the last hundred years (Butler 1998: 133). Towns may promote their contemporary attractions and new heritage experiences but they do so within the context of their spatial and cultural relationship to the Highlands: 'Gateway to the Highlands' and 'Capital of the Highlands' being phrases used to describe the towns of Stirling and Inverness respectively.

The myths of the Highlands and their literary framework are therefore still very much intact although, as noted earlier, the contemporary visitor may not realise that the Highlands they expect and see are the Highlands created for them by artistic representations. The contemporary tourist will also have their expectations created for them by the newer media of film and television. In an oft-quoted incident in 1953, described by Forsyth Hardy in *Scotland in Film*, the Hollywood producer Arthur Freed visited Scotland in order to find a suitable location for his film *Brigadoon* (1954), a film about a Highland village that wakes up for one day every hundred years. Despite being taken by Hardy around numerous villages in and out of the Highlands, Freed returned to Hollywood to film Brigadoon in the studio, declaring that 'I went to Scotland but I could find nothing that looked like Scotland' (Hardy 1990: 1). The idealised and Romantic setting that he required did not exist and so he created it in California, producing 'the archetypal film of a bogus Scotland.' As both Chapters 6 and 7 illustrate, this irony resonates with the more recent filming of Braveheart (1995) in Ireland rather than Scotland.

Later representations of Scotland on film and television have continued to concentrate on the romance of the landscape and the endearing eccentricities of the Highlanders. Popular television series such as *Dr Finlay's Casebook*, *Take the High Road*, and the more recent *Hamish Macbeth*, along with such successful films as *Highlander* (1986), *Rob Roy* (1995), *Braveheart* (1995) and *Mrs Brown* (1997), utilise the established iconography to create a sense of the Highlands that is both majestic and homely. Walter Scott's literary device of the outsider in the plot is also used to good effect in these productions. Films such as *Local Hero* (1983), for example, utilise the cynical and worldly urbanite as a way of explaining and exploring the Otherness of the local people and their connection with the landscape.

The tourism potential of film and television locations has been discussed elsewhere (Riley 1994; Schofield 1996; Tooke and Baker 1996) and Scotland has experienced this phenomenon. *Braveheart* for example has spawned a host of associated trails, attractions and theming in the Stirling area and *Mrs Brown* seems set to increase tourism in the Royal Deeside area. The Highland Council has recently set up its own Film

Commission which aims to increase film activity and consequently film-related tourism in the area. Bearing in mind that the films *Rob Roy*, *Braveheart* and *Loch Ness* created a film-related goods and services spend of approximately £12 million in the Highland area alone, it is not surprising that such initiatives have been funded.

The Scottish Tourist Board has also taken on a role in promoting this form of tourism. In the document 'Film Tourism: Business Guidelines for the Tourism Industry' (*c.* 1995), it is noted that:

> cinema is the natural successor of the Scottish literary and artistic traditions of Burns, Sir Walter Scott, Robert Louis Stevenson, Landseer's Monarch of the Glen, and Raeburn. Scotland's awe-inspiring terrain and scenery, its historical figures, literature, culture and stories have been the subject of great art, and inspired visitors then and now.
>
> (Scottish Tourist Board *c.* 1995)

It seems that the preferred imagery for Scotland's tourism has therefore remained unchanged despite the changes in media.

The real Highlands?

The Highlands, then, are partly an imaginary creation – an example of a 'place on the margin' (Shields 1991) or on the periphery of the 'real' world. In terms of development and industry, the Highlands have been seen as 'peripheral' and representing a traditional rather than a modern economy (McCrone 1996: 490–51). Womack's work on *Improvement and Romance* (1989) presents the contradiction that while the Highlands were seen as an area ripe for improvement (economic development) it was also promoted as a traditional, even feudal landscape (Womack 1989: 2–3). This dichotomy between the modern and the traditional (with the traditional imagery over-powering the very real advances made elsewhere in Scotland during the period) is often represented as part of the Scottish psyche itself and the notion of the 'divided self' (Scott 1994: 368). The Lowlands, with their centres of industry and learning, have been seen as the head of the nation and the Highlands as the heartlands. Indeed, Edinburgh as the 'Athens of the North' with its Egyptian and neo-classical architecture and monuments, can be seen as a direct antithesis to the romance of the Highlands (Grant 1988). The divided self of the heart and the head is a theme explored most famously, perhaps, in Robert Louis Stevenson's *Dr Jekyll and Mr Hyde* (1886) and William Hogg's *Confessions of a Justified Sinner* (1824). The duality of language, English and Gaelic, also adds to the construction of a divided nation unsure of its real identity amidst the myth and legend.

The continuation of tourism as a significant industry in the Highlands has meant that the region is still an imaginary playground, well-suited to the needs of the eighteenth-, nineteenth- and twentieth-century tourist:

> by leaving his home area and going to somewhere he has selected as a holiday place, the holiday-maker detaches himself completely from his normal social environment and from the ways in which he makes a living. Instead of being an expression of society, the holiday becomes a turning away from it, towards solitude, fantasy, nature, domesticity; we have already seen how closely adapted the Highland image was to the orchestration of exactly these themes.
>
> (Womack 1989: 63)

Overview

The close links between people and particular environments has been variously rationalised in terms of basic primal needs, as a physiological response to particular manifestations of landscape, or in wider socio-cultural terms. Certainly the impetus to represent and consume the world around us seems to be strong in many societies and, in the European context, the development of a landscape aesthetic has been concomitant with the rise of tourism, another means by which the environment is produced, represented and consumed. Cultural and literary motivations were an important factor in the development of European tourism and many sites on the eighteenth-century and contemporary Grand Tour were made popular and accessible by becoming the subject of literary or artistic representations.

The Highlands of Scotland are an example of one such site, a resort that was imagined into being by the twin influences of tourism and literature. An untamed and hostile terrain, peopled by war-like clans, became, through physical suppression, clearance and mythical re-imagining, a playground. This resort was a space wherein both the people and the landscape embodied the myths and legends of a simpler and entirely fictitious era.

Despite the achievements, economic development and today's parliamentary independence of this country, the myths of the Highlands remain a strong means through which the nation sees itself and is seen. The huge Scottish diaspora remembers the homeland in literature, images and song that evoke a landscape unchanged by time and images set forth by the official tourism bodies accord with this myth. The myth, of course, has its critics but it is difficult to imagine how such a successful and strongly ingrained tourism construct could be further developed to take into account the very real social and cultural change in this small northern country. R. W. Butler (1998) believes that the recognisable imagery should not be discarded in a wholesale re-branding of Scotland, and that many countries would be delighted to have such a distinctive national product.

The promotion of conference and business travel, Scottish produce and cuisine and recent developments in the electronic linking of the nation's museums and historic sites are various attempts to create a more diversified tourism product in Scotland. Hundreds of websites describing the urban and rural attractions of the country to potential and virtual visitors have been established over the last few years. A recent analysis, however, shows that by far the greatest number of websites are linked to the Highlands (Cano and Prentice 1998). Despite the creation of new markets, new products and new media, the Highland landscape and its associated mythology remain Scotland's strongest image.

6 Heritage landscapes
Merging past and present

Introduction

Chapters 4 and 8 trace the evolution of the public's interest in visiting rural landscapes and in preserving the relics of the past in urban areas. The presence of the past within the landscape itself, however, has also long evoked fascination and is a strong motivator for leisure and tourism activity. Whether a landscape has historic or cultural associations or demonstrates the marks of previous civilisations, the heritage it symbolises seems to have deep roots and suggests a wide appeal. Day trips to the countryside to visit stately homes, monuments, National Trust sites, national parks and conservation villages appear to be a popular form of heritage consumption combining the delight of the rural scene with a sense of belonging and continuity.

The idea that a sense of our history resides in the very land itself is of course not a new one. Indeed, when the National Trust was founded in January 1895 it was originally responsible predominantly for the care of landscapes under the title of 'The National Trust for Places of Historic Interest or Natural Beauty'. Registered under the Companies Act, The Trust was given the power to 'promote the permanent preservation for the benefit of the Nation of lands and tenements (including buildings) of beauty and historic interest'. *The Times* also noted that the Trust 'aims to establish a National Gallery of natural pictures' (quoted in Jenkins and James 1994: 25). It is the land and its associated buildings that represent our past and, many would argue, need to be preserved for the future.

The 'heritage guides' to Britain which appeared in the post-war period also used the appeal of landscape as the main strand in their patriotic treatment of 'our' heritage. *The Batsford Guide to England's Heritage*, first published in 1935 and reprinted in 1948 and 1961, considers the landscape in the first chapter, as does *The British Heritage*, published in 1948, which examines the 'Heritage of soil' and the 'Heritage of scenery' as priorities. The new car-owning public were thus encouraged to go out and discover their heritage in the surrounding countryside and in the monuments, farms and villages of rural England (Breese 1998: 157).

This chapter will examine the phenomenon of 'heritage landscapes' by first exploring the evolution and contemporary social construction of the heritage concept itself, identifying the meanings and problems inherent in this articulation of the past. In applying these meanings to a spatial dimension, a range of issues related to the notion of landscape as our collective heritage arises. These will be further explored by examining a range of heritage sites that challenge the notion that our past is safely preserved in a timeless and unchanging landscape.

The evolution of heritage

The evolution of the concept of heritage has, in many ways, reflected the changing attitudes to the past held by successive generations. As a social phenomenon, heritage is highly reflective of the society that has created and valued it. Heritage has also been notoriously difficult to define as various bodies and organisations have admitted (Lumley 1994: 58–9). Today, the term heritage is used in a confusingly wide arena. As a sector, it covers a wide array of protected landscapes, buildings, material culture and ways of life (Leask and Yeoman 1999). As a new form of leisure experience, heritage centres have bridged the gap between the museum and the theme park. As a branding term, it has broad marketing appeal in its associations with the past, with tradition and perhaps with craftsmanship (Thrift 1989). As a form of lifestyle consumption, it denotes a way of life associated with an elite ruling class. As a critique of post-industrial consumption, the notion of heritage as a debasing form of exploitation has had some currency (Hewison 1987). And as an appeal to particular values, it has a political dimension.

At its simplest, heritage means anything that has been inherited and this is certainly not problematic in terms of personal or family history, where a common sense of belonging and ownership is possible. Family heritage in these cases may be symbolised by ritual objects such as christening robes or wedding rings being passed down through generations. This concept of heritage becomes more controversial, however, when attempts at applying it at a regional or even national level are made. If buildings, art collections or landscapes are to represent our common heritage, questions may be asked as to what has been chosen and by whom. In a multi-cultural society, the very act of inheritance itself is problematic as the passing on of cultural symbols from generation to generation within one group will inevitably suggest a disinheritance within another. Tunbridge and Ashworth explain:

> at its simplest, all heritage is someone's heritage and therefore logically not someone else's: the original meaning of an inheritance implies the existence of disinheritance and by extension any creation of heritage from the past disinherits someone completely or partially, actively or potentially.
>
> (Tunbridge and Ashworth 1996: 21)

Given the political dimensions of this, it is no coincidence that one of the first acts of the newly elected Labour government in 1997 was to rename the Department of National Heritage to the less loaded Department for Culture, Media and Sport. A sense of national heritage had been used, it is argued, to instil patriotism and calm public disorder in its appeal to our common roots and common purpose, and more contemporary heritage manifestations such as royal events and pageantry have been used to unite the nation in times of unrest (Hewison 1987: 47). Certainly, heritage has a strong political appeal, as two consecutive Conservative Prime Ministers have discovered, with 'Victorian values' and 'back to basics' being powerful appeals to the ideals of a past era. However, it was perhaps not only the Conservative government that looked to the past in search of votes, as Sophie Breese's work exemplifies. Breese suggests that recent political rhetoric from all parties returned to inter-war patriotism in its evocation of landscape, history and the associated national characteristics of the 'English' in the run-up to the 1997 elections (Breese 1998). Such imagery may have a powerful resonance to those included in the inheritance of Englishness but may have less currency for those groups excluded from that particular heritage.

The paradigm of heritage as common inheritance is therefore problematic, as is the perceived relationship between heritage and history. It has been argued that, if history is made up of the raw facts of the past, then heritage is the processed product of history – in effect heritage is history re-shaped and made palatable for contemporary consumption. Schouten suggests that 'Heritage is not the same as history. Heritage is history processed through mythology, ideology, nationalism, local pride, romantic ideas or just plain marketing, into a commodity' (Schouten 1995: 21). This position would suggest that history is the starting-point for the heritage industry which sifts through documents and artefacts looking for attractive stories with which to beguile tourists. Thus, the selectivity of heritage may be contrasted with the undeniable evidence of history although many historians themselves would make no major claims for the truth of their discipline. All historic 'facts' have been selected and processed by some recorder in a position of privilege and each successive generation will read history from the context of their own social and cultural standpoint as the historian E. H. Carr pointed out in his series of lectures entitled 'What is History?' (Carr 1964).

Thus, the selective and socially constructed nature of heritage should not perhaps be a major facet of any critique of the phenomenon in relation to the more educational benefits of history. Spurious heritage, based on fiction and presented with the authority or at least the trappings of the museum idiom is, however, of concern (Walsh 1992: 139). Where selectivity and artistic licence are introduced it is inevitable that a dominant, and therefore possibly less problematic view of the past, will be conveyed. A heritage presentation may effectively freeze a selected, and possibly pleasing, moment from the past and prevent the visitor from gaining a sense of historic continuity.

Early commentators in the 'heritage debate' (Lumley 1994) built their arguments around the theory that there is a direct correlation between the presence of a heritage industry in a country and its economic decline. This correlation comes from the dual need to exploit the past for current economic gain and to bolster the nation in times of economic depression by an appeal to former glories – in effect – to employ nostalgia as an opiate and as a new industry. Lowenthal (1985), Wright (1985) and Hewison (1987) in their respective texts *The Past is a Foreign Country*, *On Living in an Old Country*, and *The Heritage Industry* used these arguments with much persuasion and instigated a debate on the meaning and uses of heritage which continues today. More current commentators have suggested that these earlier arguments patronise a public who, rather than simply swallowing the products of the heritage industry, actively engage with and enjoy the theatricality and pastiche of the heritage experience and are well able to differentiate this from the serious museum (Urry 1990: 100). It is also now perhaps too simplistic to consider this heterogeneous sector as a single industry (Prentice 1994: 315).

Certainly the prevalence of the past in contemporary society has been and continues to be a cause of concern and comment. In 1992, Fowler examined the huge array of 'pastness' imbued in our present day (Fowler 1992) and in the same year Walsh examined this phenomenon as a condition of postmodernity (Walsh 1992). More recently, we have seen texts produced on the management and operational aspects of different facets of the heritage industry, from heritage gardens (MacKellar Goulty 1993), to industrial heritage sites (Alfrey and Putnam 1992) as well as more discursive texts that continue to examine the meaning and use of heritage in our present times (Arnold, Davies and Ditchfield 1998). The past has become an important strand in the leisure and tourism industries, with museums, heritage centres and historic sites continuing to attract significant numbers of visitors and historic quarters and cities becoming a major part of the cultural tourism offering of Europe and beyond (Richards 1996).

The emergence of heritage as a contemporary cultural construction is therefore part of the expansion of the leisure and tourism industries, but it must also be seen within the wider context of developments in cultural production and consumption. The last three decades have witnessed what has been described as an 'aestheticisation' of our everyday lives (Featherstone 1991: 24–5). Greater access to cultural products through mass production and new technology has meant that we now consume culture in every aspect of our lives. The sharply delineated spheres of the modern period have imploded so that the boundaries between culture and commerce are now blurred (Urry 1990; Harvey 1989). Cultural events are regarded as opportunities for enterprise (Corner and Harvey 1991).

For example a recent 'blockbuster' exhibition at The Royal Academy in London displayed a huge range of associated merchandising and remained open on occasions into the early hours of the morning to meet demand. Contemporary advertising also makes much use of heritage and cultural objects, particularly classical music, to define a lifestyle and position a product.

Malraux began exploring these issues in his famous essay of 1953, 'Museum without walls'. In this work he suggested that photography in particular had been responsible for taking the consumption of culture away from the museum as institution and integrating it into everyday life, with people being empowered to create their own personal museums (Malraux 1978). New advances in technology that allow museum collections to be 'owned' and 'accessed' at home on CD-ROM (for example Microsoft's 'Gallery' which features the collections of the National Gallery) are the natural successors of the developments that prompted Malraux's work.

In terms of heritage conservation and interpretation, the social history movement of the 1970s has tempered the immediacy and aura of the authentic object itself, whether it be exhibit or building, with a greater acceptance of the wider role of the object in society. The traditional conservation of a museum object or building in its very essence demanded that it be removed or cordoned off from the world and given a new significance. Pearce describes this transforming process thus: 'the piece is lifted from the world of transient goods with its kinship with the world of the outmoded, the worn-out and the rubbish ... and into that of the culturally durable ... to be studied and displayed, protected and defended as the raw material from which cultural values will be spun' (Pearce 1994: 31).

However, the social history approach to interpretation sought to redress this 'isolation' by replacing the social and cultural context which the museum idiom removes. In effect it was an attempt to return the heritage object to society. Social historians augment objects – the hard data of history – with a range of supplementary evidence such as maps, photographs, documents and oral history, and also illustrate historic themes with objects taken from everyday experience. This allows a range of stories to be told about an object, not merely the dominant history. Thus, it is argued, the hidden stories of disadvantaged social groups, such as the working classes, women, minority ethnic groups and children can be explored (Glaser and Zenetou 1994; Hooper-Greenhill 1997).

Entire museums have been created using this museological approach, for example the People's Palace in Glasgow and The People's Story in Edinburgh, both founded in the mid-1980s by their respective Labour city councils. In these museums the histories of local industry, leisure, the experience of women and children and even the rise of local celebrities are presented. Even the latest addition to the National Museums of

Scotland, The Museum of Scotland, which opened in Edinburgh in 1998, has a contemporary gallery, where local people are invited to nominate their choice of iconic object of the twentieth century. Objects related to media, new technology, sports, health products and domestic appliances predominate in this particular gallery, where use-value seems to have replaced aesthetics in the museum choices of the Edinburgh population at least.

This process of valorising the everyday histories of ordinary people has been described by Urry as a feature of the 'anti-elitism of post-modernism' (Urry 1990: 130), where the quotidian and the vernacular are valued as much as the precious relics of church, state and monarchy. However, in terms of heritage and landscape specifically, there is evidence that this idea pre-dates the postmodern era in the manifestation of European open-air folk museums. These museums are usually credited to Artur Hazelius who opened the first of these 'living museums' in Skansen in Sweden in the 1890s. Born in Stockholm in 1833, Hazelius had travelled around much of Sweden and was passionate about Nordic culture, which he felt was being threatened by the forces of industrialisation. He founded the Nordic Museum in Stockholm and later created Skansen, a rural landscape wherein the folk practices of the rural population were rescued and preserved in the vernacular architecture, agricultural methods and even in the livestock breeds kept and displayed at the open-air museum.

The site of Skansen, positioned outside Stockholm, was intended to expand upon the collection of folk objects displayed in the Nordic Museum. In the open-air site, Hazelius brought together cottages, a Lapp camp, charcoal-burners' huts, wild and domestic animal stocks and later complete farmhouses, workshops, mills and even a working railway (Hudson 1987: 121–2). His ideal was to create a living landscape wherein folk-dancing, music and crafts would be displayed within the context of an assemblage of authentic (if transposed) buildings. G. B. Thompson describes the site as having similar aims to the Nordic Museum but that Skansen also hoped to 'bring it an enlarged dimension and, for the first time, to incorporate into the museum environment the sky, the clouds, the weather and the seasonal cycle' (quoted in Hudson 1987: 124).

The Skansen model has been transported and appears across Europe, North America and latterly Southeast Asia (Hitchcock, Stanley and King Chung 1997). Sites dedicated to recounting past rural lifestyles were opened in Wales in 1949 (St Fagan's), in Ulster in 1958 (Cultra), in England in 1967 (Singleton, Sussex) and in Scotland in 1944 (Kingussie, the Highlands). All these open-air museums feature vernacular architecture reassembled on one site and demonstrate an attempt to reconstruct the landscape around the buildings. As Hudson has noted in relation to social history, such developments are perhaps inadequate as they celebrate the lives of the rural poor with no equivalent attention being given elsewhere to the lives of the industrial worker (Hudson 1987: 126). However, in attempting to show a living landscape and the interrelation between humans and their changing environment, they were an important stage in the development of the heritage idiom with respect to the landscape.

The consumption of heritage has become an important element of our cultural life and various attempts have been made to explain its increasing significance as the above discussion exemplifies. As individuals, we find our own particular sense of our past through a range of personal objects and experiences, some provided for us by the heritage industry itself. Collectively, our heritage is demarcated, supported and presented to us by a range of bodies sponsored by local and central governments and the not-for-profit sector. More recently, the commercial sector has discovered the profits to be made from selling the past, whether in association with authentic and legitimate heritage sites or in the more spurious creation of heritage themed experiences (Schafernich 1996).

So far, this chapter has concentrated on the prevalence of heritage in our everyday lives and explored contemporary issues arising from the widespread consumption of this cultural product. In particular, the ways that different approaches to heritage interpretation can be used to arrest historical processes or to unveil them and the means by which a sense of collective heritage can exclude a wide range of potential consumers have been explored. These issues are as relevant to the heritage manifest in the landscape as they are to the urban scene or the museum collection and consequently, the discussion will now turn to heritage landscapes, their significance and their users.

Heritage in the landscape

Although a sense of heritage may be more readily conveyed through historic buildings and monuments that have a defined period and architectural style, the past and our relationship to it is also embodied in the landscape. These landscapes may display the marks of previous civilisations or they may symbolise our collective inheritance of the natural environment such as 'heritage coasts' (Tunbridge and Ashworth 1996: 2). All valued landscapes in Britain will of course have some mark of civilisation on them as the natural environment has been worked upon and exploited by successive generations, leaving remains of habitation, burial and agriculture that may only be visible to the expert eye. Such landscapes, rich in human meaning, have become known in landscape valuation/history terms as 'associative' or 'cultural' landscapes (Jacques 1995).

Recently, UNESCO have decided to add to their list of World Heritage Sites such associative landscapes, thus giving more credence to the notion that the landscape can have important historical meaning and that 'natural' areas cannot be valued in purely aesthetic terms but will inevitably have cultural associations. The International Union for the Conservation of

Nature (IUCN) was joined with ICOMOS (International Council on Monuments and Sites) to prepare criteria for evaluating landscapes that lay between the purely natural and the cultural. Jacques discusses the problems these bodies faced in applying criteria to the concept of associative landscapes of world significance and illustrates the issues inherent in this by describing the case of the Lake District, nominated as a World Heritage Site. The Lake District, undeniably a place of outstanding natural beauty was nevertheless not deemed significant in terms of glaciation and wildlife. Its significance, of course, lies mainly in its cultural associations and its place in the history of European art, literature and philosophical enquiry. After much deliberation by the organisations concerned it was decided that these associations could not be considered to be of world significance and so a decision as to the site's inclusion on the World Heritage List was deferred (Jacques 1995).

This particular case clearly illustrates an important issue inherent in the idea of heritage landscapes: that what is considered to be of heritage significance is subjective and very much temporally and contextually bound. A series of historic, social and artistic processes formed our impressions of the Lake District and the effect of these processes cause the area to be valued still in aesthetic terms (Urry 1995). But for whom is this landscape significant? As Jacques comments, 'the case for recognition of The Lake District rested to a considerable degree upon the importance of William Wordsworth and John Ruskin in nineteenth-century English landscape tastes. Yet would an Indian accept them as figures of world importance?' (Jacques 1995: 99). It is perhaps impossible to judge the meaning and significance that individuals, regions, nations or even global villagers will find in particular landscapes.

Today the greatest proportion of tourists come from urban areas and, consequently, they seek out rural landscapes as an important element of the tourist experience. Heritage embodied in these landscapes may be a prime motivator for tourist trips or may lie more deeply hidden as in the case of associative landscapes. In either circumstance, the landscape is attractive to contemporary urban visitors as it embodies a constant force in times of rapid change. Boniface and Fowler note that it is no mere accident that ' "wilderness" is described as "primeval," "landscape" as "timeless," "scenery" as "untouched," "countryside" as "unchanging". Yet of course, the essence of that greater part of the globe's landmass which is rural is precisely that it has changed and is changing all the time' (Boniface and Fowler 1993: 79). This nostalgic need to create and consume a static heritage may be widespread but must be put in the context of the prevailing theories on preservation, conservation and heritage management that have evolved over time both in relation to the built heritage and to the care of landscapes (Ashworth 1996: 166). The social history movement that created a new museology dedicated to showing historic process should also be considered as an important element in the evolution of landscape interpretation.

At the turn of the century, for example, monuments and their land-scapes were frozen at a particular point in time as a preserved relic of that moment. Emerick discusses the work of Sir Charles Peers, Chief Inspector of Ancient Monuments from 1910 to 1933 whose influential methodology of 'repair and not restore' was carried throughout Britain and the Empire. More recent approaches to heritage and landscape management allow a more fluid history to be presented – one that takes into account a variety of stories. Emerick suggests that we now take a more holistic view of the built heritage and its spatial context: 'we have rediscovered the landscape. We realise that we can only understand abbeys, priories and castles by looking at the evidence from their adjacent and distant economic hinterlands' (Emerick 1998: 188).

Thus, landscapes and their cultural associations are not phenomena to be cordoned off and preserved according to a series of universal criteria, as UNESCO discovered. What we find significant and appealing in a landscape will change over time, as will the landscape itself, as it evolves. It can be read as a palimpsest, 'a text upon which men and women have written their own social autobiography, without however, being able to erase entirely the contribution of their ancestors' (Reed 1997: xii). The range of histories written on a landscape may be the unconscious source of their appeal but inevitably these overwritten texts are also very difficult to evaluate and con-sequently manage. The cultural landscape can be prized for a wide variety of assets depending on the valuer, as Ihse and Norderhaug demonstrate in the context of Nordic cultural landscapes (Ihse and Norderhaug 1995: 159). Nature conservationists, cultural conservationists, agriculturalists, social historians and landscape architects will place very different values on land-scape features. On the other hand, the more casual visitor too will seek out different attributes and, in the case of a number of well-known heritage landscapes, the existence of a variety of histories, interpretations and users can cause management challenges or even conflict. The three following case studies illustrate heritage landscapes with very different characteristics exemplifying the major challenges in the interpretation and management of such sites in the landscape. In the example of Stonehenge, the contested elements of the site and the many voices of its users are examined. In the case of Avebury, the presence of a multitude of visible pasts within the changing landscape is explored and, finally, in the heritage landscape of Tintagel, the intangible but very potent dimension of myth within the site is discussed.

Stonehenge: multi-vocal landscape

Which story is to be the officially sanctioned, preserved and presented history? What is to remain buried? And who are the legitimate users of the heritage resource? Bender's (1993a) work on the 'contested landscape' of Stonehenge clearly illustrates the above problematic (Bender 1993a,

1998; Bender and Edmonds 1992). As a 5,000-year-old site, Stonehenge's mysterious circle of upright and lintel stones has fascinated visitors for centuries and has been the subject of many artistic representations, such as those by Turner and Constable, together with the dramatic use of the stones in Thomas Hardy's *Tess of the D'Urbervilles*. Ousby (1990: 96) describes the early tourist facilities available at the site from 1739 onwards although the first traveller to report a visit in his journal was a Swiss tourist in 1562 (Chippindale cited in Hetherington 1996: 164).

We are uncertain as to the actual date, mode of construction and, in particular, the purpose of the stones, and this mystery has left the site open to a range of interpretations. Owned by English Heritage and the National Trust, however, the site is presented and interpreted as an icon of deep national history and a monument to a shared past. This sense of a legitimate history for the site belies the fact that Stonehenge has been the centre of mythical speculation for centuries, the stones having been variously associated with Merlin the Wizard, Boadicea, King Arthur and even Robin Hood (Hetherington 1996: 163; Ousby 1990: 93). Bender outlines the very many different Stonehenges that have existed, from prehistoric to more contemporary manifestations. The site has been, in its long history, revered, feared, plundered, bought and sold and now presents itself to us as a multi-vocal and dynamic landscape, 'constantly open to renegotiation' (Bender 1993a: 276). However, she fears that the sanctioned history – that which is interpreted for the legitimate paying visitor – is history frozen in a particular moment, perhaps echoing the turn-of-the-century Sir Peers approach already outlined above:

> Stonehenge will be 'explained' in terms of roots, and of 'our' 'deep' national past. 'They' – the builders of Stonehenge – will be homogenous. It will tell the story of those empowered to make decisions and to make claims on other people's labour – a top-down history. It will be a 'frozen' past, one that ends when the last stone went up.
>
> (Bender 1993a: 270)

However, once a year, at the summer solstice in June, scenes at Stonehenge suggest that there are many other voices and alternative readings of the site, readings that include the mysticism that is part of the history of the monument. Since the turn of the century, Druids have been worshipping at Stonehenge, using it as a meeting and ritual centre. This was tolerated by the authorities as, despite their unorthodox beliefs, they were otherwise respectable tax-paying citizens. It is also suggested that their exotic robes and rituals lent them an appeal as a tourist attraction (Bender 1993a: 272). Certainly, contemporary postcards of the site with attendant Druidical worshippers are sold as part of the Stonehenge experience. In the 1970s the 'hippy' community also adopted the site as a spiritual hub, believing that psychic forces align themselves at this particular point in Wiltshire.

More recently, free festivalers and New Age travellers have made a pilgrimage to the site for yearly celebrations of the summer solstice and it is this particular social grouping, with their alleged association with drug-taking and deviant social behaviour, that prompted the custodians of Stonehenge to react with more and more intense policing of the site, culminating in the creation of exclusion zones around the stones. The climax of the tension between the free festivalers and the authorities took place in June 1985 with the Battle of the Beanfield and the arrest of 500 travellers. The final summer solstice of the last millennium saw yet another battle between the legitimate guardians of the site and alternative users for whom the stones have a very different meaning. The perimeter fence around the monument was torn down as hundreds of celebrants attempted to reclaim the site as the sun rose (*Guardian* 22 June 1999). Undoubtedly, no postcards of this heritage event will be available.

Thus, the contested nature of this particular landscape becomes very obvious as it annually transmutes into a physical and ideological battleground. Tensions between an array of users – archaeologists, stewards, historians, interpreters, day-trippers, educational groups and more alternative users – suggest that this is not a site that can be presented as a one-dimensional monument to the past. A host of histories, both scientific and mythical are embedded in the site and the challenge to represent its multi-vocality and the contested nature of both its past and its present, is very real.

Avebury: evolving landscape

Such extreme conflicts may be unique to Stonehenge but are illustrative of the notion of a multi-vocal site, perhaps less contested elsewhere but certainly problematic in terms of interpretation and management. The site of Stonehenge is a physical monument and a cultural landscape, the popular meaning of which evolves as much from eighteenth-century Romanticism and myth as from archaeology. Sharing World Heritage Site designation but less well known is the site of Avebury in north Wiltshire, 20 miles north of Stonehenge. Constructed earlier in the Neolithic period than Stonehenge, the site known as 'Avebury' actually consists of a huge prehistoric landscape incorporating long barrows, ritual sites, an extensive avenue, a ceremonial mound and a massive stone circle surrounded by a deep bank and ditch.

This huge site is to be found within a much wider prehistoric landscape, liberally dotted with Neolithic and Bronze Age sites (Malone 1989: 18). However, unlike Stonehenge, it is not set apart in isolation but has become integrated within more recent settlement. The village of Avebury has evolved over the centuries within the precincts of the huge circle itself, thus showing, in its different architectural styles, a continuity of use. The following description of the village gives some idea of this sense of continuity:

there is a bewilderment of architecture. A cottage with a buckling, dark-tiled roof looks across to a stately brick house with mansard roof and dormer windows. Near a reconstructed stone a corrugated shack stares at a row of terraced cottages of uncertain style with ornate tiles and lumpish archways that contrast with the imposing stone gateway surmounted by stone balls that lead to the manor and the pigeon-house.

(Burl 1979: 17)

There is evidence that the site has been used in a variety of ways throughout its lifetime. There are Iron Age fortifications nearby and several villas were constructed in the area during the Roman period. The Saxons used the site as a defence point and sanctuary, later building a village and church. This church was enlarged in the twelfth century and it is significant that it was sited outside the stone circle itself, giving more weight to the idea that the stones had been associated at this time with pagan cults. Popular medieval names for the stones and surrounding features, such as the Devil's Quoits and the Devil's Chair are still used today (Malone 1989: 121) and this connection between the site and devil worship was so firmly established by the thirteenth century that stones from the circle were being buried to rid the village of such pagan influence manifest in these imposing reminders. Ideas about the stones' powers changed as William Stukely informs us in his records of the site made between 1719 and 1724. Local people, keen to expand the village of Avebury, were involved in removing, burning and consequently breaking up the stones to provide building materials and this activity continued until 1825 (Malone 1989: 121–2).

Archaeological excavation was begun in Avebury in 1908 by Harold St George Gray. Interrupted by the war, significant work did not resume until 1934 when Alexander Keiller bought the whole site that by now housed garages, sheds and pigsties. He cleared and excavated the site and placed his findings in a small museum open to the public, who were also admitted onto the site itself. Significantly, the stones that had been buried to remove their pagan powers were re-erected and he demolished redundant cottages and barns found within the stone circle, a policy now discredited (Malone 1989: 123–33). The site is now in the custodianship of the National Trust and English Heritage and is interpreted through the Alexander Keiller Museum.

As with Stonehenge and other enigmatic Neolithic structures, archaeologists are uncertain as to the actual purpose of the site although clearly its massive proportions testify to its great significance. Competing theories on the societies who created the site and their motives for doing so have been put forward. Inevitably, the stones have been linked to the Druids and King Arthur in the absence of any knowledge of pre-Roman history. More contemporary theories have described the site as a ceremonial place where rituals linked to birth, death and fertility were played out by a society trying to make sense of their world. More ambitious ideas involve a society headed by learned priest figures who had developed a sophisticated understanding of geometry and astronomy (Burl 1979: 28). It has also been suggested, in a learned study, that the site of Avebury itself is in fact an image of the pregnant Mother Goddess at harvest time, created at that particular spot because of a happy topographical accident which resulted in a fusion of the living landscape and human culture:

> The Avebury district was chosen as the centre of attention because the deity was seen to function through the fortuitous alignment of two river confluences with sunrise and moonset positions at the summer quarter days. This extraordinary chance interpreted as a divine exhibition of harmony between underworld, terrestrial plane, and sky, inspired a splendid response.
>
> (Dames 1996: 10–11)

A recent visit to the site on a November morning embodies these competing theories and demonstrates the multiplicity of meanings inherent in Avebury. Among the people within the circle that morning were workmen shoring up a few almost-toppling stones with gravel, and a group of archaeologists relaying turf on an adjacent site – the past and the future of the site being attended to. A clutch of National Trust members, following their official guide books, were joined by more casual strollers, who, in the absence of much formal interpretation were letting their imaginations lead them. Such visitors are in the majority and, according to Burl: they 'come from a vague interest in the past that remains ill-defined because there is not much available to make the focus clearer. For such visitors Avebury offers an emotional response, not an informed answer to their questions' (Burl 1979: 18). And sitting on the bank was a group of alternative users of the site, quietly watching the archaeologists at work but clearly absorbed in their own less visible engagement with the forces of Avebury.

As a tourist attraction the site reflects this heterogeneity of users. Excavated materials are on display at the Alexander Keiller Museum and in the Great Barn, the Museum of Wiltshire Life sets the site within a wider context. The National Trust and English Heritage have their ubiquitous shops in the village, there is a vegetarian restaurant and souvenir shops, one of which is called 'The Henge' where visitors can buy moonstones and new-age paraphernalia along with more orthodox merchandise. The village is small and at busy times becomes overwhelmed with visitors. The custodians are aware of the problems of balancing the demands of tourists with the future of the site. However, development, expansion and agricultural practices are as much a threat to Avebury as visitors and their concomitant needs (Malone 1989: 26).

This particular heritage landscape embodies many different meanings and reflects our changing attitudes to the past. Was Alexander Keiller right to clear sites of habitation from the stone circle when their presence is testimony to an evolving cultural landscape? Currently, we believe that he was not. How do we reconcile the peaceful and pastoral site of today's Avebury with the belief that the site may have been a bustling prehistoric 'metropolis' (Dames 1996: 9)? Certainly, at the time of their intended use, both Avebury and Stonehenge would have been far more densely populated than they are now, and yet it is our contemporary wont to present these landscapes as rural and wild (Boniface and Fowler 1993: 79).

It is not easy to present a sense of evolving history or to interpret it as such, but at Avebury the different pasts and users seems peaceably to coexist. Being less well-known than Stonehenge, the prehistoric complex has escaped becoming iconic and those who wish can still roam freely among the stones. Such freedom has its drawbacks as the custodians discovered in June of this year. Pressure groups protesting against genetically modified foods daubed a number of the stones with paint, causing one in particular to resemble a huge tomato (*Guardian* 19 June 1999). The use of the monument as a site of protest demonstrates that the stones of Avebury still have a potency, albeit in the context of an act of vandalism.

Tintagel: mythical landscape

Representing historic and archaeological fact alongside more mythical dimensions is difficult in terms of scholarship. Yet these stories may be a legitimate part of the site's history and an important element in the contemporary understanding of the landscape. In describing the heritage landscape around Tintagel Castle in Cornwall, Robb discusses the challenge of managing a site that is known to most visitors through Arthurian legend and literary associations rather than through formal history or archaeology (Robb 1998).

The landscape of Tintagel is described by Robb as part of a 'heritage complex', an area where commercial activity resides alongside the official monument and where the extent of this activity is such that it may affect the visitors' experience of the heritage site itself:

> as conceived here, 'heritage complexes' include the commercial exploitation of local heritage themes adjacent to the bounds of official protection. Businesses within these complexes exploit heritage resources for profit and help form, reflect and reinforce visitor expectations.
>
> (Robb 1998: 580)

In the particular case of Tintagel, it is the mythical elements of the site that are exploited commercially, with a range of themes linked to King Arthur, Camelot and the Round Table being articulated. Merchandising, catering and accommodation in the locality centres around this imagery and a commercial visitor attraction entitled King Arthur's Great Halls orientates the visitor to the Arthurian legends through an audio-visual presentation that is a focal-point for tourist visits to the area. Tintagel Castle and Island itself makes no claim for the Arthurian legend, the site being interpreted as the home of Richard, Earl of Cornwall from 1233. However, the English Heritage shop and exhibition convey a more mixed message. The shop sells Arthurian merchandise while the display is sceptical of these very myths. Thus the visitor is torn between two messages conveyed by the official custodians of the site: 'the one voice, critical and reflective of officio-academic agnosticism on Arthur, the other drawing deeply on the legendary and literary accretions to the site in order to sell T shirts and posters' (Robb 1998: 587).

The castle and island seem to be adrift amidst legend that English Heritage will not acknowledge officially. For example, evocative landscape features popularly known as King Arthur's Seat, Hipbath and Footprint are difficult to find and not included in the official guide to the area. Robb voices concern that little attention is given to the source of the legend or in fact to competing theories on the myth of King Arthur, the use of the site over successive centuries and the basis for its identification as Camelot. Given that the visitor will have come to the castle through the pre-orientation stages of the 'heritage complex' with its Excalibur and Merlin imagery, not to mention their own bank of associated images derived from literature, television and film, some acknowledgement of this dimension should perhaps be given. It has already been seen that the social history approach to heritage interpretation attempts to re-contextualise monuments and objects and tell the lesser stories hidden by the official narrative. An awareness of the mythical dimensions of heritage landscapes should perhaps be seen as part of this movement. Stone and Mackenzie concur with this position in their work on the 'excluded past' in history education. They believe that the purely scientific interpretation denies a heritage site its contemporary relevance and strands it in 'romantic isolation' (Stone and Mackenzie 1989: 118).

Merriman's research on the public's understanding of heritage, *Beyond the Glass Case* (1991), examines what he calls 'alternative archaeology' which is proving popular in the face of official archaeology, particularly among younger people. Merriman's (1991: 113–14) survey revealed that a significant proportion of the public believe in the presence of 'mysterious forces' at sites such as Stonehenge and this desire to find meaning in belief-systems unsanctioned by science or archaeology can be seen, Merriman suggests, 'as symptomatic of the wider social phenomenon of alienation and disillusionment with science' (Merriman 1991: 116). The contemporary media has doubtless fuelled the imagination but such enduring mythic structures are founded in preliterate cultures and have their place within the landscape of heritage. If a myth is considered to be true by a large proportion of the public then:

this general belief gives it a contemporary validity. It may contain elements that are unhistorical or ahistorical but it adds up to a cultural truth. It may indeed contain a great deal of historically accurate and factually testable material, but this is transformed into a touchstone of national, local, even individual identity.

(Hewison 1989 quoted in Robb 1998: 581)

King Arthur, Druids, Robin Hood or Mother Earth Goddesses, whether they inhabited these sites or indeed existed at all, may have as much to tell us about our relationship to the past and the uses we make of the past in our contemporary society as more tangible and learned sources.

Overview

Heritage is a powerful force in contemporary society and it is articulated through our towns and cities, through the collections of material culture that we value and put on display, and through the landscape. As a deep source of meaning about our collective past and as a resort, the countryside has become especially valued in this century. The visible and the less tangible marks of past civilisations within this rural landscape have also been a major attraction for leisure and tourism activity.

The contemporary construction of heritage has been criticised as being overly selective, of hiding historic processes beneath glamourised, nostalgic presentations of the past and, in its appeal to the communal and the collective, to exclude many social groups within a multi-cultural society. However, heritage has also been seen to be anti-elitist and to accommodate our more contemporary concerns within its presentations.

Heritage embodied in the landscape is also open to these influences. Interpreters find it easier to present a specific moment in time but how can this be possible in an ever-changing landscape? A collective national history may be a desirable dimension to present in a heritage landscape of world renown but how does one reconcile such a history with a myriad of other histories being read, interpreted and acted out at the very sites themselves? And finally, how does heritage interpretation actively acknowledge the contemporary relevance of myth and alternative beliefs in the presentation of sites without resorting to trivialisation or tokenism?

As, yearly, more of our landscapes become designated World Heritage Sites, these problems, inherent in the very concept of landscape itself, will continue to present themselves. If we are to continue peacefully to live with the past, then its future should be allowed to evolve naturally and its contemporary relevance to our everyday lives must be acknowledged and celebrated.

7 Gendered landscapes
Constructing and consuming leisure and tourism

Introduction

This chapter seeks to develop many of the discussions introduced in the previous two chapters by examining the interconnections between the spatiality of leisure and tourism landscapes and the representation of these landscapes. In addition to providing synergy between the first half of the book, with its focus on 'ways of seeing' or representing landscape, and the second half of the book, with its focus on ways of consuming landscape, the discursive development presented in this chapter is intended to introduce gender as a key signifier in both the representation and consumption of leisure and tourism landscapes.

The chapter will examine the developing discourses of feminist and gender theory as these have been applied to the study of leisure, tourism and heritage landscapes. The first discourse to be discussed is that of spatialised feminism which represents a body of theory and knowledge emanating from social and cultural geography over twenty years ago. This body of research drew on socialist feminist analyses and was undoubtedly influential in informing the feminist leisure studies of the 1980s. These studies of women and leisure form the second discourse to be introduced within this chapter. As social and cultural geographies were developed further in the 1980s, they were responsive to wider developments in social and cultural theory that emphasised a shift towards gender research. The discourse of gendered space, developed by geographers but now gaining increasing credibility within leisure and tourism studies, forms the third discourse to be discussed in this chapter. Here we examine a range of insights into the gendered nature of leisure and tourism landscapes. These investigations have been undertaken more commonly in geography, cultural studies and gender studies than in leisure and tourism studies, and material is drawn from a range of disciplines and subject fields to illustrate the diversity of gendered representation and consumption in leisure and tourism. The chapter then introduces the concepts of dualisms and the Other which, together with the concept of the gaze introduced in Chapter 2, serve to illustrate the complexity of representation and consumption of leisure and tourism spaces, places and landscapes. Many of these concepts and illustrations are then brought together in a discussion of gender and landscapes of tourism. These are then illustrated further through the presentation of a case study of gender and landscapes of heritage. This discussion of gendered heritage landscapes builds on the analysis of the mythical heritage landscapes introduced in the previous two chapters.

Spatialised feminism

Within social and cultural geography there is an established literature of spatialised feminism spanning more than two decades (McDowell 1993a). Spatialised feminism refers to feminist analysis that identifies and explains the spatial dimensions of power relations between the sexes. These spatial dimensions include the differential use, control, power and domination of landscapes and can often be seen in the representation and consumption of spaces and places. The term spatialised feminism serves to encapsulate a wide range of research, undertaken between the mid-1970s and the early 1990s, which drew on structuralist analysis, and particularly socialist feminist analysis.

McDowell (1993a) identifies Burnett (1973) as producing the first paper to discuss gender issues in geography. Burnett and other contemporary researchers were largely concerned with issues related to gender and urban space or gender and 'third world' development (Garcia-Ramon, Castener and Centelles 1988; Peake 1989; McDowell 1993a, 1993b; McDowell and Sharp 1997; Women and Geography Study Group 1997). In relation to these research themes, McDowell (1993a) has stated that:

> a considerable body of knowledge was built up throughout the 1980s documenting the extent to which women were unequal and subordinate to men. Mapping the forms of this subordination and the extent of geographical variation in their (our) status in the contemporary world is perhaps the most clearly geographical project that has been undertaken by feminists in the discipline.
>
> (McDowell 1993a: 163)

It is only retrospectively, however, that the connections between these previous areas of research and leisure and tourism landscapes have been recognised and articulated. For example, Horst (1981) edited the first book by geographers on 'third world' women, making explicit the gender relations of global capitalism that had been recognised only implicitly by previous research. Horst and others therefore provided, unwittingly, a series of foundations for later feminist analyses of the gender relations of tourism as a significant component of global capitalism. Similarly, Massey (1984) added a feminist perspective to previous work that had analysed the spatial and social divisions of work and urban labour. The inter-relationships between the sexual and spatial division of labour and similar divisions visible in leisure and tourism are now apparent (Kinnaird and Hall 1994; Sinclair 1997).

Like feminist leisure studies, which were to follow in the 1980s, the spatialised feminism of the 1970s conveyed a certainty about the category 'woman' as a key social division. This homogenous vision of women has since been criticised as failing to problematise fully the complex inter-relation between gender, sexuality, spatiality and power. One source of this alleged failure was the emphasis on structural rather than cultural determinants of spatial relations, and this chapter will go on to illustrate the importance of cultural and symbolic representations of gender in leisure and tourism. The spatialised feminist research within geography relied upon distributive data to demonstrate that society's gender relations were replicated and/or mirrored in the spatial relations of housing, transport and the workplace. This approach has since been classified retrospectively, and somewhat critically, as the 'add women and stir' method, but can be seen throughout the social sciences as a significant attempt to modify a patriarchal academy.

Feminism and leisure landscapes

Within feminist leisure studies, the publication of *Women and Leisure: A State of the Art Review* by Talbot (1979) is viewed by a number of writers as representing the first piece of published research on women and leisure (Deem 1988: 7). Talbot's research formed one of a series of ten special reports funded by the Social Science Research Council (SSRC) and the Sports Council. It was accompanied by a workshop on Women and Leisure at the 1979 Leisure Studies Association Annual Conference and was closely followed by a paper on 'Women's Leisure' given by Hobson at a conference on Leisure and Social Control hosted by the Centre for Contemporary Cultural Studies (CCCS) in January 1980 (Tomlinson 1981). These events of 1979/80 can be seen as acting as a catalyst for further feminist research undertaken within the subject field over the next decade.

This chapter now proceeds to present a brief outline of the major feminist leisure research of the last twenty years. No claim is made, however, to review all the feminist literature within the field of leisure studies. Moreover, literature relating to some specific areas of the field, such as women and sport, is not addressed here as the intended focus of the review is on the dominant and emergent themes within *leisure* research.

While Stanley (1980) called for a 'radical feminist alternative' to the study of leisure, the ensuing research all tended to adopt a liberal or socialist philosophy (Deem 1986; Green, Hebron and Woodward 1987; Wimbush 1986). Moreover, aspects of leisure such as pornography and prostitution, which clearly symbolise patriarchal control, received detailed scrutiny from radical feminist analysis in women's studies but escaped attention within leisure studies (Dworkin 1981; Itzin 1992; Jeffreys 1999; MacKinnon 1995). More recently, however, a few feminist researchers working across leisure studies and gender studies have begun to draw on radical feminist critiques within a poststructural framework (Adkins 1998; Clarke 1998; Humberstone 1997). This recent work appears to offer possibilities for disrupting dominant discourses within leisure studies by stepping outside the patriarchally constrained domains of socialist and liberal feminism, while simultaneously engaging with the diversity and complexity of women's lives acknowledged by poststructuralism.

Socialist and liberal feminism provided the dominant theoretical perspectives for a number of publications based on specific case studies in the UK during the 1980s. These case studies were specific to particular groups of women, particular types of leisure participation, and to clearly defined geographical areas, spaces and places. Each case study served to highlight a further aspect of the feminist leisure agenda and to clarify particular aspects of leisure and social relations. While most of these case studies were undertaken from a socialist feminist perspective, and underpinned by a concomitant interest in class relations, there seems little doubt that they paved the way for contemporary feminist research in leisure that has now begun to adopt different theoretical perspectives.

Dixey and Talbot's (1982) study of women bingo players in Leeds served to highlight the existence of gendered leisure space, the male domination of public leisure space, and the response by women of carving out a spatial sanctuary in the form of the bingo hall. This was one of the first studies to look at women's occupation of public leisure space, as previous studies had focused upon women's leisure around the home and family and the different meanings and motivations for women participating in leisure compared with those for men (Rapoport and Rapoport 1975).

Deem's study of women's leisure in Milton Keynes (1986) stressed the relationship between access to power and access to leisure by identifying that whereas 'few women thought of themselves as having a right to leisure; many men do see themselves as having such a right' (Deem 1988: 14). Furthermore, Deem's research played an important role in beginning to establish the spatial inequity of perceived leisure opportunities for women and men. Whereas men's leisure choices were often determined by 'location, social class, money and interest', women's leisure choices were dominated by fear of violence where the desire for 'safe' leisure meant 'going where there are other women, good transport, and few or no men' (Deem 1986: 7). Henderson and Bialeschki (1991), echoing Deem, later confirmed that:

> A number of leisure researchers have indicated that women believe they have no right to leisure and this belief places a severe limitation

on their construction of time for leisure and their attitudes toward leisure (e.g., Bialeschki and Henderson 1986, Deem 1986, Govaerts 1985, Henderson 1990a, 1990b, Henderson *et al.* 1989, Wearing and Wearing 1988, Woodward and Green 1988).

(Henderson and Bialeschki 1991: 51)

Deem's research explored further some gender differences identified in Dixey and Talbot's work by highlighting the differences *between* women in relation to their access to leisure and control of public leisure space. Her work focused on the difficulties which working-class women or women with young children may experience in gaining access to leisure: difficulties which middle-class women or women without children may not experience to the same degree. Moreover, Deem began to make connections between the gendered construction of leisure and the gendered construction of power that creates both the knowledge and the practice of leisure:

> There is, however, a more general point here, concerned with the necessity of directing feminists' attention not only to the existence of power relations but also to how and why those power relations are sustained by ideological and material factors. There is much discussion about the importance, for example, of gender ideologies but little attempt to explore how these actually work in the field of leisure.

(Deem 1988: 14)

The heterogeneity among women, and of women's leisure opportunities and experiences, was further identified in a study of leisure participation by women with pre-school age children undertaken by Wimbush (1986) in two distinct areas of Edinburgh. The study, funded by the Health Promotion Research Trust, attempted to 'evaluate the importance and meaning that mothers attach to their existing leisure opportunities' and 'to explore the ways in which social and recreational activities feature in their general health and well-being' (Health Promotion Research Trust 1987: 2). In meeting these objectives, the research identified and evaluated a range of factors that both constrained and facilitated women's involvement in leisure. The recommendations for leisure policy and practice outlined by Wimbush serve to summarise the emphasis and ethos of feminist intervention in leisure from the mid-1980s to the mid-1990s. The eight recommendations listed focus on the need for structural change in the provision of public leisure services and were later adopted and adapted by other leisure providers in addition to being reiterated in later feminist leisure research (Aitchison 1997; Aitchison 2000d; Aitchison, Brackenridge and Jordan 1999; Aitchison, Jordan and Brackenridge 1999). Wimbush's recommendations consisted of:

- play and creche facilities as an integral part of more social and recreational venues, so that children are well looked after and mothers are freed from making complicated childcare arrangements;
- places where children are welcomed and where they can participate in activities with other children or with their parents;
- settings where women feel comfortable and not threatened or out of place (women-only sessions in sports, leisure and adult education centres taught or led by women are one way of putting women at ease and increasing their self-confidence);
- activities for women in the evening as well as during the daytime, so that participation can be fitted within the busy timetables of employed mothers as well as those working at home;
- flexible payment systems that allow participants to pay-as-they-go, so that those not able to afford block payments are not prevented from taking part, and those unable to attend regularly are not financially penalised;
- a means whereby providers and policy-makers can consult regularly with women in the community (e.g., via locally based 'outreach' workers) and learn from the self-help arrangements that some women have developed to meet their leisure needs;
- information about leisure activities through the local media and via nurseries, clinics, health centres, playgroups and shopping centres.

(Health Promotion Research Trust 1987: 2)

The largest and most frequently cited piece of research, however, was that conducted in Sheffield by Green, Hebron and Woodward who undertook an SSRC and Sports Council sponsored research project to investigate *Leisure and Gender: A Study of Women's Leisure Experiences.* This research, published in 1987, later formed the basis of a comprehensive research text on the subject of women's leisure (Green, Hebron and Woodward 1990). The Sheffield research confirmed many of the findings from the Edinburgh project and 'indicated that women with children under the age of five are least likely to feel satisfied with the amount of free time they have' (Woodward, Green and Hebron 1988: 137). The research, however, went wider in its exploration of 'the social processes through which women's access to leisure is constructed, defined and regulated' (Woodward, Green and Hebron 1988: 144). Green, Hebron and Woodward (1987) drew on both socialist and radical feminist theory to explain women's experiences of leisure. Ultimately, however, their recommendations for implementing practical change in the short-term relied on liberal feminist intervention directed at policy-makers. Drawing on socialist feminist analysis, Green, Hebron and Woodward (1987) point to women's lack of disposable income as a result of their labour market position in capitalist society. Their acknowledgement of radical feminist theory can be seen in relation to the importance they attach to the patriarchal control of women's time and space, particularly in relation to private or domestic space. This focus on the dual system of capitalist and patriarchal oppression is highlighted by Woodward, Green and Hebron (1988), before emphasising that change may have to be the result of less radical and less far reaching transformation to public leisure policy and practice:

> The best prospect for promoting women's recreational opportunities seems, in the light of this work, to be in promoting greater autonomy and power for women within relationships and in the wider society, so that these discriminatory attitudes and behaviour become anachronistic and socially unacceptable. In the meantime, policy-makers within leisure and recreation may need to 'work round' the effects of such inequalities, in relation to the pricing and time-tabling of use of their facilities, the development of women-only sessions, and the introduction of a much wider scale of transportation to and from leisure venues.
>
> (Woodward, Green and Hebron 1988: 145)

Overall, however, the influence of socialist feminism is evident throughout the major case studies outlined above. Wearing (1996), drawing on poststructuralist theory, offers a critique of these previous feminist leisure studies:

> Emphasis in sociological theory on the centrality of the labour process as explanation for the workings of capitalist society has obscured the importance of other spheres of life in the everyday lives of individuals. Feminist theorists have challenged the male bias in such theories, but still have concentrated on the labour aspects of the family as explanation.
>
> (Wearing 1996: 186)

In contrast to contemporary social and cultural geographies of leisure, which have stressed the importance of everyday consumption in constructing gender relations, there seems little doubt that previous feminist leisure studies have viewed class relations as a major determinant of leisure relations. For example, Woodward, Green and Hebron (1988: 99) list six determinants of leisure with the first three emphasising the influence of social class: 'the most significant influences on women's leisure were their social class, level of household and personal income, employment status, age group, marital status and stage in the family life cycle'.

While these feminist studies of leisure undoubtedly disrupted the discourse of leisure studies, they can also be viewed as part of the sociological tradition of leisure studies where, 'from the very beginning, leisure research in Europe was a topic oriented field of research dominated by

sociological perspectives but strongly leaning toward public policy interest'
(Momaas, 1997: 241). In this way,

> much theorising and research in leisure takes place within a norma-
> tive citizenship paradigm which regards public leisure provision as a
> central component of social citizenship and largely ignores the predom-
> inantly commercial nature of modern leisure. This combines with a
> reproductionist, society in leisure, approach to emphasise leisure as
> an area of inequality and inequity with a consequent downplaying of
> issues relating to the nature of leisure experiences.
>
> (Coalter 1998: 21)

Throughout the leisure studies literature there are attempts to define
leisure in terms of 'time', 'activity', 'experience', and 'space' (Haywood,
Kew and Bramham 1989; Wearing 1996). All of the case studies outlined
above contain a wealth of information about the gendered construction
of leisure time, activity, experience and space. Spatiality, however, remains
largely untheorised as a result of the dominant emphasis on social class
issues and the influence of feminist sociology rather than feminist geography upon leisure studies in the 1980s. Similarly, structural analysis rather
than cultural analysis is foregrounded within these research projects: 'the
leisure studies orientation is toward broader social aggregates and issues
of inequality of power, hegemonic processes, collective identities, access
to and provision of opportunities, and social citizenship' (Coalter 1997:
257).

Consequently, socialist feminist analyses have also had the effect of
rendering invisibile sexuality in the research, both as an identity form and
as a process or product of cultural relations. As this chapter will illustrate, both spatiality and sexuality in leisure relations are reintroduced by
the research undertaken within recent social and cultural geographies
which have given greater priority to cultural analysis in addressing issues
of the spatiality of gender and leisure. Unlike feminist leisure studies, the
cultural turn in geography has taken discourse as its starting-point in an
attempt to reconfigure power relations within the sub-discipline, seeing a
reworked discourse as a prerequisite to a redefined hegemony and ideology
within geography.

Gendered space

The development of a focus on gendered space rather than spatialised
feminism can be seen as an attempt to accommodate many of the criti-
cisms levelled at the structuralist feminist approaches outlined above.
Acknowledgement of the role of gender-power relations in shaping spatial
relations has turned our attention from the recognition of inequality to
the recognition of difference. This perspective can be identified as another

of the post-positivist discourses identified by Barnes and Gregory (1997: 2) and introduced in Chapter 2. Together, the perspectives of poststructuralism, post-colonialism, postmodernism and poststructural feminism can be seen as providing a challenge to the homogenising and totalising category 'woman'. Indeed, Haraway has stated:

> There is nothing about being 'female' that naturally binds women. There is not even such a state as 'being' female, itself a highly complex category constructed in contested sexual scientific discourses and other social practices. Gender, race or class consciousness is an achievement forced on us by the terrible historical experience of the contradictory social realities of patriarchy, colonialism, racism and capitalism.
>
> (Haraway 1985: 191)

Theorising gendered space offers acknowledgement that gender-spatial relations, which may manifest themselves in the representation and consumption of spaces and places, are contextually, temporally and locally specific. However, a number of feminist theorists have cautioned against the wholesale adoption of a poststructuralist perspective as such contextually specific analyses negate theories of power as systemic phenomena. With no systemic power relations there can be no overall system of domination and oppression, only specific contexts of subordination, resistance and transformation. One of the questions raised in this chapter, then, is the extent to which systemic male power – that is, patriarchy – exists in relation to the construction and consumption of landscape, and/or the extent to which localised, contextualised and pluralised power relations exert their influence on the spatiality of gender.

In relation to leisure, Rojek (1995: 1) has emphasised the importance of poststructuralist analysis to our understanding of leisure in late modernity, stating that, 'one cannot separate leisure from the rest of life and claim that it has unique laws . . . the object of leisure is subsumed by the subject of culture'. In an appraisal of Rojek's (1993) earlier work, however, Coalter acknowledges that 'postmodernism has produced a crisis in leisure studies' collectivist theorising and has undermined the ability of most leisure studies scholars to sustain their commitment to emancipatory politics' (Coalter 1997: 259).

This feeling of crisis can also be identified in feminist leisure research as the theoretical developments made in the 1980s, and measured in the form of such tangible outputs as feminist leisure texts (Deem 1986; Green, Hebron and Woodward 1990; Wimbush and Talbot 1988), do not appear to have been matched by similar progress in the UK during the 1990s. Scraton has attributed part of this crisis to 'the new malestream world of theory that is engaging with postmodernity debates':

leisure research in the postmodern world appears to focus on conspicuous consumption – the public world, the tourist gaze, hyperrealities of Disney World, theme parks, heritage, etc. There is a need to put the personal and political back on the research agenda. Feminist research has always rejected the political separation of the public and the private. It has highlighted the significance of the private world for women, a world that is highly gendered. Leisure research has begun to open this silenced, private world by recognising the importance of the private sphere in the control of women's leisure (Deem 1986; Green *et al.* 1987). Issues of time, space, sexuality, regulation, control, and power remain important in the 1990s and need to be further explored. By turning our leisure gaze onto the public arena the gendered reality of the hidden private world becomes once more invisible and ignored.

(Scraton 1994: 259)

In spite of this critique of postmodernism, Scraton's later work acknowledges that Foucault's 'notion of power as ubiquitous and fluid rather than harnessed has some value' to feminist leisure studies (Watson and Scraton 1998: 126). Further, it is argued here that poststructuralist studies of gender and leisure conducted within gender studies, cultural studies and social and cultural geography have offered alternative ways of seeing and theorising 'time, space, sexuality, regulation, control and power' within gendered leisure relations (Aitchison 2000e).

In stark contrast to the concern expressed by Watson and Scraton (1998: 125) that poststructuralism has offered 'a predominantly male gaze onto the public arena', Green (1998: 173) states that 'the postmodern concern with diversity and difference has opened up areas of "private" life previously neglected by key social theorists in the 1970s and 1980s'. In addition, two significant advances during the 1990s, which have contributed to feminist and gender leisure research, can be seen as symbolic of a wider restructuring of feminist and gender analysis. First, within the UK, feminist leisure studies have been adopted and adapted by other disciplines and subject fields in a move which is representative of the increasing breakdown of disciplinary boundaries within the academy. Second, advances in feminist and gender research in other countries with an established record of leisure research, such as the US, Canada, Australia and New Zealand, have had an impact upon UK feminist and gender leisure research. As Henderson (1994: 129) has emphasised 'new understandings of power may help to present a framework that transcends the dualism of sex differences by presenting the complexity of powerlessness which some women and other "disenleisured" groups confront in their daily lives'. Both of the developments outlined above have engaged with poststructural theory in ways that leisure studies in the UK has been reluctant to encourage to date. Wearing (1996) comments that:

In most poststructuralist feminist analysis there is a shift from thinking about 'patriarchy', with its emphasis on male control of the structures of society, to the notion of 'phallocentrism' which denotes male control of language, symbols, definitions, discourses, sexuality, theory and logocentric thinking.

(Wearing 1996: 39)

J. Butler (1990) sees poststructural feminism as providing a challenge not just to phallocentrism but also to compulsory heterosexuality which, when combined with phallocentrism, enables the continuation of male domination through both structures *and* cultures. Increasing leisure analyses undertaken within other subject fields and disciplines in the UK, such as social and cultural geography, sociology, gender studies and cultural studies, have drawn extensively on Butler's work, on poststructuralist theory in general, and have been instrumental in developing critiques of binary divides and dualistic thinking in social theory (Aitchison 2000e; Aitchison and Jordan 1998b; Wearing 1996). Moreover, research from the US and Australia has been influential in bringing poststructural analyses of power, constraint, empowerment and resistance to studies of leisure and gender (Henderson 1994; Henderson and Bialeschki 1991; Henderson *et al.* 1989; Henderson *et al.* 1996; Samdhal and Jekubovich 1997; Wearing 1996).

Deconstructing dualisms

Many of the liberal and socialist feminist studies of women and leisure referred to above have highlighted the problems of defining leisure and of undertaking empirical research of women's leisure. Since the rise of public sector leisure provision in Victorian Britain, formal leisure policy, provision and participation, in relation to both built facilities and more open landscapes, has been defined and dominated by men. Industrial capitalism, and its distinction and separation of work and leisure, public and private, production and reproduction, led to what Momaas (1997) has described as 'gender-specific time-spatial segregation'. The associated set of binary oppositions has inevitably defined leisure in relation to paid (male) employment.

Within all of the major schools of thought positioned within leisure studies, theorisation of leisure has taken place in relation to work. Drawing on Wilensky (1964), Parker's (1971, 1983) conventional leisure theory defined leisure as an 'extension' or 'opposition' to work or as 'neutral' in relation to work. Parker (1988: 11) later discussed revisiting his earlier definitions to consider the possibility of adding a fourth 'category of "exploitation"; that is, in our own lives or in our society we can exploit work for leisure'. Given the often-exploitative nature of gender relations within both work and leisure, this addition could have served to strengthen further the feminist project within leisure studies. Roberts' (1978) pluralist leisure theory emphasises the influence of work upon leisure, while simultaneously warning against exaggerating the significance of work which, he argues, is but one of a number of social variables which act together to determine leisure meanings, motivations and behaviours. Together, Parker and Roberts have been referred to as the 'Founding Fathers' of leisure studies:

> Despite a growing recognition of leisure as a sub-discipline within sociology, studies in the area are still relatively marginalised and often ignored altogether in general textbooks, or are included as a subsection of chapters on paid work. The 'Founding Fathers' of the sociology of leisure in Britain emerged with the publication of their textbooks on leisure in the early 1970s, most notably Stanley Parker and Kenneth Roberts. Before that, students and researchers in the area were limited to American material of a functionalist nature (for example Gross, 1961) or philosophical writings on the relationship between human nature and the need to play (Huzinga, 1950). Exceptions to this lay in the area of empirical sociology, a good example of which are the British community case studies, which concentrated on traditional working-class communities (Young and Willmott, 1957; Dennis, Henriques and Slaughter, 1969). Such studies examined the nature of everyday life, emphasising the close social interaction within the community, and leisure was depicted as a vital part of community life.
>
> (Green, Hebron and Woodward 1990: 11–12)

Interestingly, this emphasis on the everyday was later to form a strong focus of the poststructural social and spatial analyses of leisure carried out under the umbrella of contemporary social and cultural geographies and discussed in Chapter 2. However, the neo-Marxist analysis of leisure developed by Clarke and Critcher (1985) continued to refer to the centrality of work, as constructed by industrial capitalism, in determining leisure patterns. Where gender differences and inequalities were acknowledged, these were attributed to capitalist oppression and its associated sexual division of labour rather than patriarchal oppression. Even studies investigating women's 'leisure lives' have analysed leisure in relation to paid employment (Scraton 1994). For example, Kay (1996: 49), in 'Women's work and women's worth: the leisure implications of women's changing employment patterns', examined 'the interrelationship of work and leisure in British women's lives'.

The binary divide of work/leisure has therefore served to marginalise many informal leisure activities from the leisure research agenda. Simultaneously, our gaze has been directed away from everyday landscapes and towards the more formal landscapes associated with paid employment and *re-creation* from paid employment. 'Hidden' forms of leisure associated with the home, with children, or related to household work, shopping, or everyday consumption are therefore frequently omitted from empirical research within mainstream leisure studies. In contrast, these informal leisure forms have begun to be researched and theorised within social and cultural geography, cultural studies and gender studies where binary divides have been challenged and deconstructed within post-structural analysis. Notwithstanding these theoretical developments, incomplete definitions of leisure, which have operated within traditional leisure studies, have served to perpetuate gendered definitions of leisure and gendered perceptions of leisure. In addition, and as Rojek (1988: 14) has emphasised, this traditional definition and analysis of leisure 'focuses on the socially approved forms of leisure ... they do not talk about deviance or deviant activities in leisure ... to my mind their work has a very impoverished view of power relations'.

But perhaps the binary divide of sex/gender, identified by J. Butler (1990), is the dualistic construct which has given rise to most debate within poststructural feminism and has now begun to be acknowledged as significant within leisure studies (Aitchison 2000e; Wearing 1996). As Wearing (1996: 47) points out: 'It is, however, a reflection of the present day male-defined dichotomy of mind/male, female/body that sociological studies to date, even those with a feminist emphasis, have generally concentrated on women's bodies and men's minds'. The distinction between sex and gender has, therefore, been criticised as yet another dualism that has served to elevate the male sex at the expense of the female gender:

> To expose the foundational categories of sex, gender, and desire as effects of a specific formation of power requires a form of critical inquiry that Foucault, reformulating Nietzsche, designates as 'genealogy'. A genealogical critique refuses to search for the origins of gender, the inner truth of female desire, a genuine or authentic sexual identity that repression has kept from view; rather, genealogy investigates the political stakes in designating as an origin and cause those identity categories that are in fact the effects of institutions, practices, discourses with multiple and diffuse points of origin. The task of this inquiry is to centre on – and decentre – such defining institutions: phallogocentrism and compulsory heterosexuality.
>
> (J. Butler 1990: viii–ix)

Within feminist geography, the gendered construction of dualisms and binary oppositions has been revealed in relation to the dichotomies of nature/culture, urban/rural, masculine/feminine (Rose 1993). The post-Enlightenment valorisation of such dichotomies within western society is founded on, and subsequently reproduces, constructions of the Other.

The gendered Other

Rose (1995: 116) explains the process of constructing the Other or Othering as 'defining where you belong through a contrast with other places, or who you are through a contrast with other people'. Characterised by dualisms, this process inevitably defines norms and deviants, centres and margins, cores and peripheries, the powerful and the powerless. Rose (1996) later drew on the work of Irigary to illustrate that the binary distinctions between self and other, and real and imagined space, are part of what J. Butler (1990: 13) has referred to as 'the epistemological, onto-logical and logical structures of a masculinist signifying economy'. Thus, the construction of dualisms or binary opposites is inherently related to the construction of the Other. Drawing on the work of Cixous (1983), it is possible to identify three fundamental relationships within this process. First, the construction of the Other is dependent upon a simultaneous construction of 'the Same', or something from which to be Other to. Second, this relationship is one of power whereby that which is defined as 'Same' is accorded greater power and status than that which is defined as Other. Third, that which is defined as Other is accorded a gender and this gender is always female.

Both feminist and anthropological analyses have informed the theoretical conceptualisation of Othering (Clifford and Marcus 1986; de Beauvoir 1949; Fowler and Hardesty 1994; Said 1978, 1993; Wilkinson and Kitzinger 1996). In both cases conceptualisations have relied upon cultural rather than structural analyses and, within feminist psychology, Wilkinson and Kitzinger (1996) present one of the most comprehensive accounts of Othering in relation to gendered power:

> In contemporary feminism, then, the concept of women as Other (sometimes with a capital 'O', sometimes not; sometimes in scare quotes, sometimes not; sometimes italicised, more often not) involves the central claim that Otherness is projected on to women by, and in the interests of, men, such that we are constructed as inferior and abnormal.
>
> (Wilkinson and Kitzinger 1996: 3–4)

Within tourism, the creation of a unique place or tourist destination frequently employs the social construction of the Other. This has been commented upon at length within the literature of tourism anthropology and tourism sociology, and there is widespread agreement that tourists often engage in what Van den Berge (1994) has described as 'The Quest for the Other' (Urry 1990; MacCannell 1992). The reverse side of the coin in searching for the Other, however, is searching for 'the self'. While the search for the authentic Other may lead to the discovery of the authentic self, Brown (1996: 38) has argued that 'the tourist seeks out the inauthentic Other in the quest for the authentic Self'. Brown's view provides a role for the tourist which builds upon both Boorstin's (1964) and MacCannell's (1992) critiques of tourism as a 'pseudo-event' or as 'staged authenticity' and gives recognition to tourists as a heterogeneous group with different 'cultural capital' (Bourdieu 1986).

The attempt to create and market places based on specific tourist attractions or a unique type of heritage must coexist, however, with the need to facilitate the tourist's identification with such a place. This echoes the point made in Chapter 2 concerning the pre-requisite need for a commonality of interpretation by those who gaze upon monuments in order to construct a dominant cultural narrative. Thus, a balance must be sought between presenting something which is different and presenting something sufficiently familiar for the tourist to identify with. This can be achieved by using vehicles such as exoticism, eroticism, nationalism and militarism, combined with ancestral heritage, to provide the connections between the present and the past, near and distant, the familiar and the Other. Most writers in tourism anthropology and tourism sociology cite Said (1978) as the originator of the concept of the Other in reference to the Western social construction of 'Orientalism' (Selwyn 1996). The anthropology of tourism therefore discusses the concept of the Other in relation to anthropological representations of race and ethnicity and this male-dominated discourse then ignores the wealth of feminist writing on the Other underpinned by the much earlier references to woman as the Other by de Beauvoir (1949):

> Humanity is male and man defines woman not in herself but as relative to him; she is not regarded as an autonomous being ... She is simply what man decrees ... She is defined and differentiated with reference to man and not he with reference to her; she is the incidental, the unessential as opposed to the essential. He is the Subject, he is the Absolute – she is the Other.
>
> (de Beauvoir 1949: 18)

In relation to the discourse of Othering, post-colonial feminist theory has drawn upon and developed many of the tenets of poststructuralist and feminist theory. Brooks (1997: 105) has commented that 'this intersection is an interesting one, as all three movements have in common the process of dismantling or subverting hegemonic discourses'. Such subversion, she argues, is achieved through the displacement of dominant discourses by marginal epistemologies. This process involves engagement and dialogue with dominant discourses rather than dichotomous opposition. Moreover, like postmodernism's relation to modernity, post-colonialism can be seen as signifying 'a critical engagement with colonialism, not to claim that colonialism has been overturned' (Spoonley 1995: 49). A discussion of post-colonial feminist theory provides a useful reference point from which to look back at historical constructions of women as Other. Mills (1998: 98–9) contends that post-colonial feminist theory has had three distinct impacts. First, by focusing on the global rather than the Western parochial, post-colonial feminism has challenged the authority and representativeness of white feminist theorising. Second, post-colonial feminism has challenged the masculinity of post-colonial theory and its lack of engagement with the dynamics of gendered power in colonial and imperial contexts (McClintock 1995). Third, rather than merely providing deconstructionist critiques of previous feminist theory or post-colonial theory, post-colonial feminist theorising has begun to build theory in its own right. Here, the work of Spivak (1985, 1987, 1993) has been instrumental in questioning white Western feminists' representations of 'Third World women' as Other through the homogenising of the category 'women'. As we stated in Chapter 2, this is what Memmi (1967: 85), writing about 'the colonised', described as 'the mark of the plural' (all Others look the same) and is related to Mead's original concept of 'the generalised Other' (Mead 1934). Like the post-colonial theories of Said (1978, 1993), Spivak and other post-colonial theorists have paid particular attention to the role of language and discourse in creating and sustaining power relations.

Such theory enables us to reflect critically upon discursive constructions of leisure and tourism destinations as foreign and exotic, and the people within these landscapes as Other. For example, the ethnographic postcard or the tourist brochure purports to illustrate that which is both indigenous and authentic (Edwards 1996). But, as Enloe (1989: 19) has emphasised, 'selecting postcards is one of those seemingly innocent acts that has become fraught with ideological risks'. These photographic images (re)construct both people and landscapes out of context, while simultaneously projecting a particular symbolism in the way in which artefacts and events are pictorialised. Invariably, such images, whether they be of people (usually women) or places, convey impressions of exotic, unspoilt, natural, virginal and desirable spaces. These landscapes, including corporeal landscapes, are frequently represented as the canvas upon which the explorer or the tourist can make his mark by consuming Other cultures. Building on Bourdieu's (1984) concept of cultural capital, it has been acknowledged previously that the medium through which the tourist destination is consumed contributes to our cultural capital of travel-related knowledge (Urry 1990). Thus, post-colonial discourse has begun to render possible a critique of the legacy of colonialism and its associated cultural capital represented in contemporary leisure and tourism.

Gender and landscapes of tourism

Travel and tourism have become associated with a globalised 'melting pot' where postmodern deconstruction and reconstruction have induced the breakdown of previous national, cultural and geographical bound-aries. Paradoxically, however, the mechanisms employed in contemporary tourism development have simultaneously served to strengthen, rather than destabilise, gendered representations of space and place alongside notions of nationalism and bounded cultural identities:

> Host societies differentiated by race/ethnicity, colonial past, or social position from the consumer societies are sold with feminised images. The tourism product – as a combination of services, culture, and geographic location – is consumed in situ, in various transactions from tourist gazing to the selling of otherness.
>
> (Swain 1995: 249)

In both the creation of the exotic and the erotic, gender plays a central role in the construction, representation and consumption of tourism land-scapes. A gender analysis of tourism landscapes therefore has to address issues related to the social and cultural construction of space and place in addition to expanding existing analyses of tourism provision, employ-ment and management. In short, tourism needs to be considered not just as a type of business or management but as a powerful cultural form and process that both shapes and is shaped by gendered constructions of space, place, nation and culture. Moreover, as Swain (1995: 247) has empha-sised: 'tourism, as leisured travel and the industry that supports it, is built of human relations, and thus impacts and is impacted by global and local gender relations'.

Although tourism is frequently cited as 'the world's fastest-growing industry' or 'the world's largest industry' it is less frequently identified as the world's most sex-segregated industry or the world's most sex-role-stereotyped industry. A growing body of tourism literature has analysed the gendered construction of employment and management within tourism (Kinnaird and Hall 1994; Sinclair 1997). There is little research, however, identifying the gendered representation, production and consumption of tourism landscapes (Aitchison 1999b, 2000c). Much of the previous research addressing issues of gender and tourism has been criticised for adopting a dualistic perspective on hosts and guests which, according to Wearing and Wearing (1996), creates a false dichotomy only too remi-niscent of the patriarchal object/subject relations discussed above. Instead of adopting this rather structuralist approach which suggests hosts are people who have tourism *done* to them by guests, Wearing and Wearing stress that tourism relations are far more complex, dynamic and dialectic. Their use of poststructuralist theory to analyse tourism's interrelations provides a helpful route to understanding the negotiations and transgres-sions which take place within, and as a result of, tourist encounters, both between people and between people and places. This perspective begins to shed light on the simultaneous existence of both freedom and constraint in the same tourist experience or in the same location.

The idea of tourism as liminality, or as freedom from the gendered constraints of everyday life, is evident in the writings of the women travellers of the nineteenth and early twentieth centuries. Privileged by class and race, many of these women undertook journeys and expeditions unthinkable by today's women supported by all the trappings of modern tourism (Blunt and Rose 1994; Mills 1991, 1996). These journeys, however, were also subject to enormous constraints and difficulties, many of which were gender-related (Blake 1990). In relation to contemporary tourism, Gibson and Jordan (1998), in their cross-national research of solo women tourists from Britain and the United States, have demonstrated that women use a number of complex strategies to enhance their safety while travelling. These strategies echo the findings of Valentine (1989) and are largely concerned with the construction of safe space through the adoption of particular travel patterns and transport choices or the construction of particular mental maps of spaces, places and landscapes which they feel they should avoid when travelling alone. Research conducted by Aitchison and Reeves (1998), and reported in more detail in Chapter 9, concurs with these findings and demonstrates that women, and especially women identifying as 'lesbian', employ elaborate environmental interpretation techniques in their attempts to travel in safety. In addition, this theme of constraints in women's tourism patterns also extends to their tourism roles. Both Davidson (1996) and Deem (1996) have demonstrated that tourism behaviour often continues and perpetuates the gendered domestic behaviour of families and heterosexual couples where the gendered roles of the home are merely transposed to another location.

But while women adopt particular gendered roles as tourists or are reified and objectified within postcards, tourist brochures and sex tourism, we are frequently invisibilised within heritage tourism. In contrast to other forms of tourism, the heritage landscape often emphasises nationalism and the construction of the nation-state by drawing upon a history that frequently renders women the invisible Other. Here women can be absorbed, subsumed and finally rendered invisible by engendered representations of nation.

Gender and landscapes of heritage

Associations between places and their history have been reconstructed by what Hewison (1987) has termed the 'heritage industry'. As Chapter 6 illustrated, this has resulted in the creation of 'a whole new breed of attractions and intermediaries who supply culture specifically for tourist consumption' (Richards 1996: 13). In Britain, this new heritage industry has made connections between specific places and traditional industries in the case of Beamish in the former coal-mining area of north-east England and Ironbridge in the former iron smelting area of the English–Welsh border. Similarly, English literary figures have been used to create heritage attractions in the form of Brontë Country in Yorkshire, Jane Austen Country in Hampshire, and Thomas Hardy Country in Dorset. Even associations with semi-mythical figures in English history have been used to market landscapes such as Robin Hood Country in Nottinghamshire. In spite of these developments, and a concomitant and burgeoning interest in heritage studies, the role of gender in the construction of heritage landscapes seems yet to capture the imagination. So although heritage has become a major preoccupation of leisure and tourism studies, there has been a lack of acknowledgement of the role of gender in both the representation and consumption of heritage places and products. Whereas Chapter 2 demonstrated that geographical discourses have engaged with the concept of gendered space in the creation of place and landscape, leisure, tourism and heritage studies are only beginning to debate issues of gender and spatialisation (Aitchison 1999a; Aitchison and Jordan 1998; Watson and Scraton 1998).

This section attempts to bridge the divide between contemporary social and cultural geographies and leisure, tourism and heritage studies by offering a critique of gendered representations of heritage and their role in the creation of gendered spaces, places and landscapes within cultural tourism. In doing so an attempt is made to contextualise and engender heritage by teasing out the multiplicity of interconnections between space, place and gender experienced in relation to the social and cultural construction of heritage landscapes and iconographies. Instead of viewing heritage as merely occupying material space, the symbolic nature of heritage landscapes and the significance of 'the gaze' and 'the Other' are foregrounded in a discussion of engendered heritage representation and consumption. A specific case study of a heritage tourism landscape is used to contextualise this discussion. The case study builds on the work of Edensor and Kothari (1994) by extending their analysis of the 'masculinisation of Stirling's heritage' through a more detailed conceptual and empirical analysis of the way in which heritage tourism is represented within the Scottish town of Stirling.

While Stirling's tourism industry has undoubtedly benefited from the recent films of *Braveheart*, *The Bruce* and *Rob Roy*, it has relied upon particular forms of masculinism, militarism and nationalism to create gendered spaces, places and landscapes. These gendered places are created and contested through the representation and interpretation of the iconography of a heritage landscape which emphasises masculine visibility and superiority in landmarks and monuments, buildings and statues, signs, symbols and banners. In addition, the content and discourse of postcards and promotional literature provides textual material that supports particular readings of the town's material culture.

Edensor and Kothari (1994: 165) argue that heritage production and consumption are gendered through a series of processes which, 'articulate masculinised notions of place and identity, and male dominated versions of the past which privilege white, male, heterosexual experience and activity'. In relation to Stirling, they go on to identify the three sites of the Argyll and Sutherland Highlanders Museum at Stirling Castle, the Bannockburn Monument and the Wallace Monument as forming the 'Stirling Triangle', with each point reflecting, 'particular and partial histories and myths, male-defined landscapes, and gendered national identities'. Although these three sites are undoubtedly influential in constructing Stirling's gendered heritage they are also part of a wider landscape of landmarks and monuments that represent additional gendered identities while simultaneously mirroring and reasserting the power of the 'Stirling Triangle'. The three towering landmarks consist of: the castle which is located at the highest point of the old town; the monument dedicated to William Wallace and situated on an escarpment to the north-east of the town; and the memorial to Robert the Bruce located on a hill to the south-west of the town. All represent performances of gender and power which are replicated both materially and symbolically throughout the town. Moreover, the tripartite message of nationalism, militarism and masculinism, performed by these three monuments, is reinforced by the portrayal of additional male icons and visual imagery drawn both from fact and fiction, history and heritage.

Ashworth (1994: 16) has stressed that 'history is the remembered record of the past: heritage is a contemporary commodity purposefully created to satisfy contemporary consumption. One becomes the other through a process of commodification'. In the case of Stirling, this process of landscape commodification has attempted to create an identity of place that distinguishes the town and surrounding countryside from competing destinations. Traditionally, this identity has been constructed around the image of Stirling as 'The Gateway to the Highlands'. But such an identity implies a short-stay tourist destination through which visitors pass *en route* to their final destination. In order to encourage longer stays and greater income generation, Loch Lomond, Stirling and the Trossachs Tourist Board, together with Stirling District Council, attempted to market the area as 'Royal Stirling' and 'Historic Stirling' during the late 1980s and early 1990s. Promotional literature developed since the late 1980s emphasises the area's links with Robert the Bruce, William Wallace and Rob Roy MacGregor, alongside frequent references to Stirling Castle as a residence of the Stuart Kings. In 1995 Scotland's forty-two local tourist boards were rationalised into a system more akin to that of England and Wales where fewer, larger tourist boards attempt to achieve greater economies of scale. But the heterogeneity of the landscape and tourist attractions in and around Stirling has created difficulty in defining one overall identity or brand label that encapsulates the diversity of the product. In an attempt to reflect this heterogeneity of place identity, the new tourist board has the rather unwieldy name of 'Argyll, the Isles, Loch Lomond, Stirling and the Trossachs'.

The production of the Hollywood blockbuster, *Braveheart*, could not have been timed better for the new tourist board and the new local authority of Stirling Council, created in April 1996. The film offered a brand image with which to encapsulate the history and geography of the area, together with the dominant discourses of nationalism, militarism and masculinism that had captured the popular imagination following the success of the film. Swain (1995: 248) has emphasised that tourism practices reflect representational issues of identity and nationhood in the marketing and consuming process performed between hosts and guests. In the summer of 1996 this interaction between hosts and guests took on a revised form as Stirling experienced increased visitor numbers and the local tourist board launched a television marketing campaign with a nationalist flavour emphasising that 'Stirling is *Braveheart* Country'. Sharp (1996: 98) alerts us to the idea that 'the symbols of nationalism are not gender neutral but in enforcing a national norm, they implicitly construct a set of gendered norms'. Penrose (1993: 29) too, comments that the construction of the category 'nation' is dependent upon the social construction of people, places and a mystical bond between people and places. This mystical bond is certainly evident in Stirling, if somewhat different from the exotic female host and the Western male guest referred to earlier in this chapter. Ironically, and in contrast to recognised profiles of tourists and visitors to heritage sites, many of Stirling's new visitors were young Scottish men with a reasserted interest in Scottish nationalism motivated by *Braveheart* and the exploits of an Australian actor (Mel Gibson) filmed largely in Ireland.

The heritage landscape of Stirling is a product of both physical and cultural geography. Stirling is situated at the narrowest point between east and west in the whole of the British Isles and also overlooks the Carse of Forth which forms the divide between the Highlands of Scotland to the north and the Lowlands to the south. The town itself grew down the side of one of three volcanic outcrops in the area which rose above the marshland of the Carse: a deep impenetrable moss which was not drained until the eighteenth century.

> Merchants, travellers, politicians and armies were compelled to seek the only main road, paved in parts, and the only bridge. In order to do so, they had to go right through the town of Stirling for the moss and marshes washed the very base of its rock. That simple

topographical fact explains why so many of the significant battles in
Scottish history were fought near to Stirling.

(McKean 1985: 3)

Stirling's economic and industrial history is an interwoven mix of market
and garrison town supported by the primary industries of coal and agriculture. With the demise of coal mining and its other related secondary
industries in the 1980s, the town focused upon financial and retail services,
its university and tourism as its major sources of income. There appears
to be a sense in which the town has sought to replace the masculinity of
the collapsed male dominated industries with forms of masculinity
embodied in Stirling's nationalistic and militaristic history. William
Wallace, Robert the Bruce, Rob Roy McGregor and the Stuart Kings
feature heavily in the town's product development, place promotion and
marketing literature. The significance of gender in informing Stirling's
tourism promotion cannot be overstated. As Sharp (1996: 103) contends,
'gender cannot be teased out of other relations of power which consti-
tute individual subjectivity but must be seen to exist contingently in all
situations – no one can be without gender and in most social locations
this is a powerful aspect of subjectivity'.

A number of ambitious tourism plans were proposed in the mid-1980s
by the then Labour-controlled local authority under the umbrella of
'Futureworld'. These plans included the building of a funicular railway
up the side of the castle rock, extensive car parking, and futuristic transport links to Stirling's other two major tourist attractions – the Wallace
Monument and the Bannockburn Monument. There was extensive local
opposition to the proposed developments, particularly from conservation
groups and the affluent residential area of the Kings Park located to the
western foot of the castle rock. This early alliance between the green
movement and the middle classes was partially responsible for a revised
series of tourism proposals which resulted in a number of developments
during the 1990s. These included: the renovation, for tourism-related
purposes, of many of the buildings in the old town; traffic management
schemes throughout the town, including work to ease the flow of traffic
in the old town which was notoriously congested by tour buses in the
summer; and the redevelopment of an extensive series of footpaths around
the castle rock. These paths were previously inaccessible to most tourists
but now form routes for scenic walks and guided tours. New developments also included extensive marketing of Stirling's nationalistic and
militaristic history with William Wallace, Robert the Bruce, Rob Roy
McGregor and the Stuart Kings forming the new iconography of the
Stirling landscape.

The representations of Wallace and Bruce within the town, as well as
overlooking the town from the two surrounding hillsides, take a number

of forms. Each male icon has a street and a number of public houses named after him. There are large statues of William Wallace at two of the most focal points within the town. The first is on the frontage of a public building called the Athenaeum which is situated at the top of King Street in the centre of the town and overlooking the main shopping area. The second is less than 100 metres away on the front of the municipal building (the town hall or local council offices). There is a large statue of Robert the Bruce in front of the castle, overlooking the castle esplanade and the view over to the Wallace Monument some two miles away.

In addition, the logo of Futureworld, the local tourism development and promotion project from the 1980s, consisted of Robert the Bruce on horseback surrounded by a rainbow. The logo of Stirling District Council, which became part of Stirling Council following local government reorganisation in April 1996, had swapped the rainbow for a banner which enveloped Bruce with the words 'Royal Stirling' above a shield displaying a picture of Stirling Castle and the words 'Key to the Kingdom'. This slogan could be seen as a reference to Stirling as the historic capital of Scotland, as the entrance to the Highlands, and as the locus of the history of the making of Scotland as a nation. In addition, local postcards feature numerous images of the two icons of Wallace and Bruce from both near and far, conveying the heroic guardianship of both the town and the nation.

Stirling's male-dominated and male-defined history is evident in the names of buildings and their previous uses. Many of the buildings of the old town are now used as tourist attractions or as restaurants and coffee shops, primarily intended for tourists. In 1994 the Stirling Heritage Trail opened featuring many of the buildings of the old town but without any critique of the gendered nature of the town's heritage or of the representation of that heritage. Old town buildings named after their original male residents include John Cowane's House, Glenngarry Ludging, Spittal's House, Darnley's House, Norrie's House, Auchenbowie's House, Mar's Wark, the Argyll's Ludging, and the Erskine Church.

In addition to residential buildings and churches Stirling's new heritage trail, which starts with 'a soldier's view' from the castle esplanade, includes many other buildings with male histories: the Old Military Prison, the Old Grammar School for Boys, the Old High School, the Tolbooth and Prison, the Mercat Cross, and the Boy's Club. So while a tour of such heritage attractions might inform the tourist about some of their forefathers there is no information about our foremothers as women are invisibilised from the local domestic history as well as from national political history and heritage.

The symbiotic relationship between masculinism and nationalism has been well documented (Yuval Davis and Anthias 1989; Enloe 1989). Less frequently discussed, however, is the role that nationalism plays within tourism promotion (McCrone, Morris and Kiely 1995). A variety of forms of media have combined to reassert Scottish nationalism in recent years and films such as *Braveheart*, *The Bruce*, and *Rob Roy* have acted as catalysts for a tourism industry desperately in search of the familiar Other.

The Wallace Monument epitomises the interrelation between masculinity, militarism and nationalism in heritage production. The monument was refurbished to incorporate a variety of new visual displays in time to benefit from the additional tourists generated by the film *Braveheart*. The 220-foot tower overlooks Stirling and the River Forth and was originally completed in 1869 as a monument to William Wallace who had led the defeat of Edward I of England at the Battle of Stirling Bridge in 1297. In interpreting such heritage Ashworth, however, stresses that:

> there is no national heritage product but an almost infinite variety of heritages, each created for the requirements of specific customer groups; viewed from the side of the customer, each individual necessarily determines the constitution of each unique heritage product at the moment of consumption.
>
> (Ashworth 1994: 15–16)

Ashworth's interpretations of the heritage producer and consumer are diametrically opposed to those of Hewison (1987). Drawing on the work of the Frankfurt School, Hewison focused on the hegemonic economic, social and cultural power of the producer and the naiveté of the unquestioning and undifferentiated mass consumer. While a feminist analysis of Ashworth's work might draw attention to the over-emphasis on individual agency, feminist critique of Hewison's work might point to the lack of recognition of hegemonic masculinism as a dominant power-broker in the representation and interpretation of heritage. Rose (1993) encapsulates these sentiments by stating:

> Whether written or painted, grown or built, a landscape's meanings draw on the cultural codes of the society for which it was made. These codes are embedded in social power structures, and theorization of the relationship between culture and society by these new cultural geographers has so far drawn on the humanist Marxist tradition of Antonio Gramsci, Raymond Williams, E. P. Thompson and John Berger. All of these authors see the material and symbolic dimensions of the production and reproduction of society as inextricably intertwined.
>
> (Rose 1993: 89)

In the work of the humanist Marxist theorists cited by Rose, landscape is constructed as part of a capitalist hegemony where patriarchy is subsumed within an analysis of capitalist power structures. It could be argued, however, that heritage has the potential to represent the reproduction of a past patriarchal hegemony by a present engendered iconography so that there is a twofold displacement of women spanning history and heritage and merging fact with fiction.

In the case of Stirling, this iconography of nationalism is also inseparable from that of militarism in the representation of Stirling's heritage. Wallace and Bruce are seen as national heroes because of their military victories. Stirling Castle and its regimental museum are tourist attractions largely because of their militaristic heritage. Complementing the statue of Robert the Bruce overlooking the castle esplanade is another statue of a soldier from the local regiment, the Argyll and Sutherland Highlanders, depicted with his bayonet drawn during the Boer War. There is a further sword-bearing statue in the form of Rob Roy MacGregor positioned at another focal point on the edge of the old town, and The Old Military Prison or Town Jail reinforces the militaristic theme further. Stirling's Old Military Prison opened in April 1996 as a combination of tourist attraction and office space for local businesses. The redevelopment took over three years to complete and cost £2.6 million of public money with funding coming from Forth Valley Enterprise, Stirling District Council, and the European Regional Development Fund. The attraction is modelled, to a large extent, on Inverary Jail whose 1996 promotional literature stated that the attraction had won a number of awards and recommendations for its displays of 'torture, death and damnation: the story of Scottish crime and punishment 1500–1700' featuring 'an introductory exhibition with blood curdling details of mediaeval punishments'. In Stirling, visitors are introduced to life-like wax models of male prisoners and guards and an audio system conveys the sounds of prisoners in agony during and after torture.

Such representations of history result in a heritage that appears to embrace forms of masculinist myth-making in an attempt to promote tourism. In addition, and in common with many other urban tourism developments in Britain, Stirling has focused upon ghost walks to attract and entertain tourists. These heritage performances are guided tours conducted in the evening and on foot with a number of additional costumed characters appearing *en route*. Similarly, Rob Roy MacGregor has had a heroic past constructed for him through postcards, statues and a visitor centre where tourists are encouraged to make up their own minds as to whether Rob Roy was a 'hero or villain?'. As Chapter 5 illustrated in relation to other narratives of Scottish 'history', the language used to describe the less than heroic aspects of the life and character of Rob Roy provides few challenges to the types of masculinism, violence and nationalism which are frequently constructed through a romanticised heritage.

Overview

This chapter has sought to contextualise and engender tourism and heritage landscapes by teasing out the multiplicity of interconnections between space, place and gender. Constructions of the gaze and the Other were introduced as central concepts in the process of tourism and heritage representation and consumption. Stirling then acted as a case study illustrating the embodied nature of the gaze and the Other within the interconnected mechanisms of place promotion, landscape construction, and mythical representation. Our engendered heritage has been identified as a complex cultural product that is continually negotiated and contested. Landscapes of heritage tourism offer gendered and mythologised representations of power that can serve to merge history with heritage and fact with fiction, whether that power be masculinist, nationalist, militarist, or a combination of all three. As Kappeler (1986) has stated:

> Representations are not just a matter of mirrors, reflections, key-holes. Somebody is making them and somebody is looking at them, through a complex array of means and conventions. Nor do representations simply exist on canvas, in books, on photographic paper or on screens: they have a continued existence in reality as objects of exchange; they have a genesis in material production. They are more 'real' than the reality they are said to reflect. All of these factors somehow straddle the commonsense divide between fiction and fact, fantasy and reality.
>
> (Kappeler 1986: 3)

8 Retrophilia and the urban landscape

Reinterpreting the city

Introduction

Towns and cities cannot escape the need to modernise: to rebuild, adapt and refurbish ageing urban landscapes, or else stagnate. People and firms migrate, new activities seek accommodation and functions change, but how to manage the transition and deal with the 'valued legacy of the past' are challenging problems for urban authorities (Tiesdell, Oc and Heath 1996). The chance survivals of earlier phases of development, in the form of medieval walls and towers, historic buildings and street patterns, may acquire special significance to the onlooker. Selected features become objects of interest, if not veneration, and a case is made to save them from demolition or unsympathetic alteration. For insiders who live and work in the locality, they are landmarks that may convey a reassuring sense of belonging to a particular place with a distinct identity and permanence. For outsiders whose commitment is short term, they offer a pleasing backdrop for business and leisure activities. Moreover, some historic quarters deemed rich in 'heritage' attain distinction as places of cultural consumption: urban landscapes which become visitor attractions in their own right.

Value systems which elevate considerations of art, aesthetics and history have frequently come into conflict with the exploitation of land and property for capital gain (Larkham 1996). For over a hundred years preservationists have found themselves embattled in dispute with landowners, developers and other agents of change. Legislation to protect a small number of 'ancient monuments' emerged in the late nineteenth century, but statutory protection for buildings of historic or architectural interest was limited in scope and the legal bargaining position of preservationists remained weak until the post-war era. In the 1950s and 1960s, voluntary organisations campaigned with vigour and determination to save historic quarters threatened by schemes for comprehensive redevelopment. In time their call was answered and, in the wake of legislation passed in late 1960s, local authorities began to formulate policies and plans for area-based conservation.

As discussed in Chapter 6, initiatives to conserve and enhance the character of old and run-down areas are now warmly endorsed by local politicians. With increasing sophistication, heritage is used to cultivate place-image and attract inward investment: an urban renaissance to improve the environment and quality of life for local inhabitants and visitors alike. Conservation has thus become more closely linked with strategies for regeneration. The Urban Task Force (1999), chaired by Lord Rogers, stresses the worth of historic buildings and townscapes as 'important assets', and concludes:

> Conservation bodies are playing an increasingly pivotal role as catalysts for urban regeneration. The expertise of these bodies needs to be exploited to the full, providing them with opportunities to contribute proactively to the renewal process, in partnership with the communities they seek to represent.
>
> (Urban Task Force 1999: 251)

The apparent consensus in recent times may, however, hide a number of important issues and problems. Choices must be made concerning which elements of the urban landscape are to be conserved and interpreted and whose heritage is represented (Ashworth 1988, 1994). Pandering to the dream-images of prospective tourists can sometimes result in stage-managed presentations of a nostalgic past that never was. Real places become themed experiences, indistinguishable from heritage centres. The nation may, perhaps, be in the process of re-imaging itself as the site of one vast open-air museum, its economy dependent on the manufacture of heritage (Hewison 1987). Harvey (1989b) construes a darker side of heritage-led regeneration, which uses illusion to disguise the intensifying class and racial tension behind the carnival mask. Places engage in programmes of aesthetic reconstruction: 'City streets have become pedestrianized and "traditional" ironwork street furniture, resonant of Victorian England, "re"-introduced. Buildings are floodlit, lending the night-time streetscape a theatrical atmosphere, and their frontages historicized and conservation policy extended' (Hughes 1998: 21–2).

Some argue that indiscriminate protection of the past stifles the creativity of the present thereby compromising the future. Writing in the *Observer*, Meades (1997) advocates an iconoclastic approach in deference to the new millennium:

> demolition on symbolic rather than utile grounds, in order to lighten the load of history rather than free the space for new buildings – though that would be a peripheral benefit. There was a time when the old was unassailable because of what might replace it. Today the old is unassailable merely because it is old. Buildings are protected because they have achieved longevity, not because they are any good.
>
> (Meades 1997: 21)

This chapter traces the origins of retrophilia, a fond attachment to historic features of the built environment. It discusses the ways in which the voluntary organisations of a heritage 'movement' have gained influence over public policy, and the processes through which they have negotiated the boundaries of what the state deems worthy of protection. It discusses the concept of townscape and its association with the idea of local or civic pride, as well as the promotion of townscapes and streetscapes as desirable settings for leisure and tourism. Reference is made to a case study of a socially disadvantaged area of inner London, where the recent designation of 'cultural quarters' and interpretation of the hitherto undervalued legacy of multiculturalism in the historic urban landscape, raises some important questions concerning the relationship between insiders and outsiders and between host communities and visitors.

Antiquity, restoration and fake

It was noted in Chapter 3 that topographers of the seventeenth century felt it appropriate to include descriptions of ancient remains in their geographies of the British Isles. To the sons of the nobility and gentry who had undertaken the European Grand Tour, these generally bore poor comparison with the more impressive sites of classical civilisation. Nevertheless, by the early eighteenth century, a select circle of dilettantes, led by William Stukeley, Roger Gale and their Itinerant Society were conducting pioneering archaeological surveys and investigations. Their particular interest was the Roman period, but prehistoric, Anglo-Saxon, and medieval remains were also explored. By the late eighteenth century, there was a much wider interest in 'Gothic' antiquities, for the rise of Picturesque tourism coincided with a new taste for native traditions (Andrews 1989). Encouraged by the writings of Gilpin, there was a quest for ancient ruins whose crumbling arches inspired poetic reverie. Historic sites with aesthetic appeal became significant tourist attractions, their images reproduced in numerous sketches, paintings and prints. Thus, the early development of archaeology, with its scientific interpretation of the material culture of the past, coexisted with the more popular leisure activity of visiting ancient remains that pleased the eye.

Some came to believe that particular sites had importance as tangible reminders of the nation's history, but little could be done if a landowner decided to demolish or alter them in a Philistine manner. In 1786, Gilpin argued that owners of property might be answerable to 'the court of taste':

> The refined code of this court does not consider an elegant ruin as a man's property, on which he may exercise at will the irregular sallies of a wanton imagination: but as a deposit, of which he is only the

guardian, for the amusement and admiration of posterity – a ruin is a sacred thing.

(quoted in Ousby 1990: 116)

The new taste for the medieval was also expressed in replication. The construction of new buildings in the Gothic style was a manifest reaction against neo-classicism and its formal rules. Replication began with sham ruins that came into fashion as decorative features of English landscaped gardens from the 1720s. By the 1760s, Horace Walpole was entertaining guests in his whimsical neo-Gothic 'plaything house' at Strawberry Hill, near Twickenham. Over the next hundred years or so, the Gothic style was to be elevated from a novel and agreeable plaything to the very embodiment of architectural purity (Ousby 1990: 106).

An early apologist for the uplifting qualities of Gothic architecture, Augustus Pugin saw spiritual goodliness, not only in individual buildings, but in the medieval urban landscapes that he illustrated in imaginative reconstructions, juxtaposed with bleak views of the alienating and degenerate modern city. The resulting paired pictures are reminiscent of the before and after imagery of early advertising, except that here the moral judgement is reversed: his 'after' represents a moral and aesthetic regression, that is meant to lead us back to a better past (Graham-Dixon 1996: 172). For Pugin, a convert to Roman Catholicism, the medieval city embodied the ordered and purposeful society that flourished before the Reformation: the antithesis of what he saw as the spiritual impoverishment of contemporary urban life. In *Contrasts: Or a Parallel between the Noble Edifices of the Middle Ages and Corresponding Buildings of the Present Day, Shewing the Present Decay of Taste*, Pugin held that the art of building churches 'like all that was produced by zeal or art in ancient days, has dwindled down into a mere trade ... They are erected by men who ponder between a mortgage, a railroad, or a chapel as the best investment of their money' (Pugin 1836: 49–50). Pugin's passion for the architecture of the Middle Ages was not simply to commend it as one type or pattern from which a modern architect might choose in a repertoire of styles. Far more fundamentally, he considered it the physical expression of the religious doctrine in which he fervently believed (Williams 1961: 138).

The relationships between aesthetics, morality and society were further explored by John Ruskin who saw Art, not as illusionist representation, but as an embodiment and expression of a universal Truth. Williams (1961: 144) highlights Ruskin's radical criticism of nineteenth-century society, reflected in its art and architecture which lacked the 'organic' qualities that he deemed wholesome. As Cosgrove (1998: 248–9) comments, Ruskin admired the truthfulness of vernacular and medieval buildings that drew inspiration from handiwork of the Supreme Architect: from natural landscape and the curving lines of flora and fauna. In *The Seven Lamps of Architecture* (1849), he paid homage to fourteenth-century Venice where he found a correspondence between the harmonious structure of the built environment – from humble dwellings to the great palaces of the merchant patriciate – and a social hierarchy that he considered just and stable. In the same work, Ruskin expressed a purist and highly impractical view of the sacred qualities of ancient buildings. These should be protected: '[w]e have no right to touch them. They are not ours. They belong partly to those who built them, and partly to all the generations of mankind that follow us. The dead still have their right in them' (quoted in Larkham 1996: 12).

Deeply conservative though his moral and social philosophy might have been, Ruskin had a significant influence on the socialist ideals of William Morris. Leading exponent of the Arts and Crafts movement, which sought to revive the ideal of handcrafted objects, Morris was 'fairly steeped in medievalism' and declared his hatred of modern civilisation. Like Ruskin, Morris thought that the architecture of the Middle Ages revealed a collectivist spirit absent from the Victorian capitalist ethic. But, unlike Ruskin, he believed this would only be regained in a future socialist society (Lowe and Goyder 1983: 22). It is therefore something of an irony that the neo-Gothic style became the dominant form for the monumental architecture of an industrial age. Leading Victorian architects used the new construction techniques of mass production, but adapted medieval symbolism to create imposing superstructures and decoration. Gothic references were used for government, civic, ecclesiastical, and significant commercial buildings, exemplified in Sir Gilbert Scott's (1873) St Pancras Station and Midland Grand Hotel. Victorian politicians and businessmen appropriated medieval forms to legitimise their quest for hegemony (Walsh 1992). As yet, however, few expressed regret for the loss of genuine pre-industrial sites. The urban landscape was being transformed by triumphal engineering projects in the form of railways, new roads, and buildings that required deep foundations and destroyed the ancient remains beneath.

Reverence, worldliness and action

It was only in the last quarter of the nineteenth century that nostalgic sentiment was galvanised into action through an organised movement to protect places of historic or architectural interest. A small but influential group of concerned individuals expressed horror at the loss, not only through urban renewal and demolition, but through insensitive alteration of ancient buildings to please Victorian bourgeois taste. In 1877, Morris was moved to establish a Society for the Protection of Ancient Buildings (SPAB) on learning that Sir Gilbert Scott, not content with plundering Gothic motifs for new buildings, had 'restored' Tewkesbury Minster. In a letter to the *Athenaeum*, Morris highlighted the need for an association 'for the purpose of watching over and protecting these relics, which, scanty as they are now become, are still wonderful treasures' (Briggs 1962: 81). The mood of optimism and belief in boundless prosperity that had char-acterised mid-nineteenth-century Britain was gone. Anxiety over the vicissitudes of the national economy, heightened by the great depression of the 1880s, was matched by a more equivocal attitude towards industry and progress (Lowe and Goyder 1983: 19–20). Tangible relics of previous epochs provided constants in a changing world. They could stand for continuity, stability and tradition, against the rootless stirrings of indus-trial capitalism: the antithesis of artificial creations and vulgar materialism in the modern age.

In the metropolis, for example, the Society for Photographing Relics of Old London was formed (1875) to record places that seemed most under threat, especially seventeenth-century houses and inns (Richardson 1995: 273). Later, the Committee for the Survey of the Memorials of Greater London (1894) was established by Charles Ashbee, a leading figure in the Arts and Crafts movement and a disciple of Morris. Its aim was 'to watch and register what still remains of beautiful or historic work in Greater London and to bring such influence to bear from time to time as shall save it from destruction or lead to its utilisation for public purpose'. The Survey Committee began a record of buildings that they deemed signifi-cant. They also published scholarly monographs, the first of which was concerned with a group of seventeenth-century alms houses in Mile End, East London that was due to be demolished (Ross 1991: 13). The survey was initially financed by voluntary contribution, but intervention by the London County Council later secured public funding and, in 1898, the Council obtained powers to spend money on the preservation of places of architectural or historic interest.

Legislation that affected the established privileges of those who owned land and property proved more controversial. The 1882 Ancient Monument Protection Act was a first step. Radical in concept but limited in scope, it gave statutory protection from demolition to fifty sites iden-tified in a schedule. The Royal Commission on the Historical Monuments of England was set up in 1908 to make an inventory of sites 'connected with or illustrative of the contemporary culture, civilization and condi-tions of life of the people of England ... from the earliest times to the year 1700 and to specify those which seem most worthy of preservation'. A subsequent Act of 1913 extended the power to purchase ancient monu-ments (without compulsion) to all local authorities and the Commissioner of Works. The Act also established the principle of 'guardianship', whereby an ancient monument could be taken into the care of a public body without the necessity of purchase. It was, however, the voluntary sector that provided the main focus for practical conservation activity and, as previously discussed in Chapter 4, the foundation of the National Trust for Places of Historic Interest or Natural Beauty in 1895 was a significant landmark. Octavia Hill (1905), a key driving force within the organisation, expressed confidence that its leadership would consist of men and women 'free from the tendency to sacrifice such treasures to mercenary considerations, or to vulgarise them in accordance with popular cries' (quoted in Lowe and Goyder 1983: 20).

The formation of voluntary agencies and the commitment of public-spirited leaders was also the basis for a radical vision of how cities should develop. As a new century dawned, the Garden City Association was set up with an agenda to marry social reform with the creation of healthier living space. Based on the ideas of Ebenezer Howard (1898), the aim was to create a series of free-standing settlements, through investment by limited-dividend companies. These would be set in relatively open country on major transport routes, but far enough from established cities to utilise cheap land. Industrialists would relocate factories and house their workers so as to combine a semi-rural environment with the infrastructure and amenities of urban centres. The aims were forward looking, but the design of the Garden City, by architects Parker and Unwin in the early 1900s, was strongly influenced by the Arts and Crafts movement. References to a preindustrial age were strong, not only in the cottage-style dwellings, but also in site plans that featured village greens, as well as the integration of woodland, streams and other natural features (Creese 1966: 169–70; Hall 1988: 86–135; Miller 1981: 74). The city of tomorrow would be built in the image of a medieval English village.

Modernism, collective memory and amnesia

In the early years of the twentieth century, however, arguments emerged for a clean break with the past and all its physical trappings. In the ideological turmoil in Europe that followed the First World War, manifestos for a new social order demanded new structures for urban living. Swiss born architect-planner Le Corbusier was, perhaps, the most famous exponent of the Modern movement. His principles for architecture and city planning, set out in *The Radiant City* (1933) and other polemic writings, envisaged high-rise, high-density living in 'cells' within mass-produced, uniform house-machines. These monolithic towers would be set in a landscape of open parkland linked by a grid of transport arteries for a new motorised society. Within the central core and among office blocks, cultural and entertainment facilities would be rebuilt for the elite. For the lower orders, accommodated in satellite units on the urban periphery, there would be generous green space and sports and leisure provision. Reconstruction would be on a heroic scale, requiring the wholesale demolition of old quarters and starting again with a clean slate. A few historic monuments would be retained *in situ* or relocated and taken out of context and displayed as *objets d'art* in the new city.

In more recent times, critics of Modernism have expressed profound relief that Le Corbusier's totalitarian plans to rebuild cities such as Paris were never implemented. Commenting on the idea of preserving a handful of historic buildings as isolated monuments, Mumford (1968) notes that his vision of the future city completely overlooked the fact 'that no small part of their value and meaning would disappear, once they were cut off from the multitudinous activities and associations that surrounded them' (Mumford 1968: 120). For those who live and work in a city, such features of the built environment become imbued with personal meaning, linking memories with the present (Lowenthal 1982). Rossi (1982) has taken this critique further in his thesis that urban landscape should be regarded as 'the locus of collective memory': the soul of a city that can survive in historic features, landmarks and street names. As Chapter 2 suggested in relation to the monument and the spectacle, these landmarks provide essential clues through which a city can be understood and represented, orally as well as in maps and other visual images. Their loss, through redevelopment or neglect, may induce a state of collective amnesia.

In the United Kingdom, the clean lines and functionalism espoused by the Modernists had a fashionable appeal to the avant-garde, and influenced a generation of young British architects with a distaste for the 'style-mongering historicism' of the nineteenth century (Casson 1975: 220). In general, however, the Modernists failed to make a positive impression on public opinion. During the inter-war years over four million new dwellings were built but urban centres tended to expand horizontally rather then vertically. As building society mortgages became available to middle-class households, home ownership increased from just 10 per cent in 1914 to 32 per cent in 1939, and suburban development spread along transport corridors into the countryside. The advertising posters of building societies seldom portrayed the reality of semi-detached suburbia, but gave accurate expression to the dreams and aspirations of home-owners. Advertising imagery alluded to the 'traditional' English cottage and garden, with the symbolic representation of threshold, hearth, and red tiled roof that have acquired the patina of age. The nuclear family was a central feature and Gold and Gold (1994) draw our attention to images of a small girl playing happily and safely in an environment which is healthy and secure.

With increased mobility by train, coach and private car, the inter-war period saw a boom in domestic tourism, and historic towns and cities became popular destinations. Publishers responded to the vogue and helped to stimulate it through well-illustrated, popular guidebooks aimed at a middle-class audience and especially those discovering the joys of leisure motoring. Most notably, there was the Batsford British Heritage series, that included titles such as *The Old Towns of England*, *The Old Inns of England*, and *The Spirit of London*. John Betjeman, future Poet Laureate, published a series of accessible and influential articles in *Architectural Review*, the *Spectator* and elsewhere, reflecting the author's fondness for Georgian and early Victorian architecture. And Betjeman reached an even wider audience through the *Shell County Guides*, which he founded in 1933. As well as the guidebook *genre*, popular themes of neo-Romantic prose included the championing of historic urban areas threatened by worldly materialism, as in Hilaire Belloc's essay 'The Crooked Streets':

> Why do they pull down and do away with the Crooked Streets, I wonder, which are my delight, and hurt no man living? ... It ought to be enough, surely, to drive the great broad ways which commerce needs and which are the life-channels of a modern city, without destroying all the history and all the humanity in between: the islands of the past. For note you, the Crooked Streets are packed with human experience and reflect in a lively manner all the chances and misfortunes and expectations and domesticity and wonderment of men.
>
> (cited in W. Williams 1942: 226)

The 1932 Town and Country Planning Act was a policy response, of sorts, to the argument for preservation. For the first time, protection was extended to occupied buildings as well as ancient monuments. Local Authorities were now empowered to make preservation orders for buildings with architectural or historic merit, but the legislation was timid, the procedures cumbersome, and it could entail compensation to owners. Parliament was reluctant to interfere with the rights of property, and little use was made of the Act (Ross 1991: 19). Controversial schemes that demolished architectural heritage captured media attention, and examples included the loss of John Nash's Regent Street to make way for new retail development, and Regency terraces in Park Lane for hotels and apartments. The cause of preservation continued to depend on lobbying, the occasional gesture by a rich person, and the vagaries of local authority intervention (Richardson 1995).

Paradoxically, in view of their intent, the Luftwaffe's 'Baedeker raids' on Britain's cultural heritage in the early 1940s provided the greatest spur to survey, evaluate and protect the nation's historic buildings. In the aftermath of enemy bombardment, the government sponsored an extensive photographic survey and an inventory of that which remained. Clive Rouse wrote in his preface to the 1943 edition of *The Old Towns of England* that many of the streets and buildings described in his pre-war guide had been destroyed:

> The destruction wrought by German bombs and fire in such places as Canterbury, Norwich, York, Exeter, Bath, Coventry, Southampton and Bristol ... is little short of a national disaster. The National Buildings Record under Mr. Walter Godfrey, though started fifty years

too late, is doing what it can to record our architectural heritage. But it must be confessed that war-time German havoc is in some cases no worse than the destruction caused by ignorance, callousness and criminal apathy of local authorities. One hopes that the loss of so much may make people all the more appreciative of what is left and more zealous in its protection ... Let us hope that wise preservation and restoration and far-seeing post-war planning may do something in the future to counteract the blow.

(Rouse 1943: Preface)

In the post-war era, the resolve of policy-makers to create a better Britain had been bolstered by the adverse effects of pre-war laissez-faire, as well as the devastation wrought by the war itself. The 1947 Town and Country Planning Act enshrined the key principles of comprehensive planning through control of development. With regard to buildings of special architectural or historic interest, the Minister now had a statutory duty to compile a list that would guide local authorities in carrying out their planning functions. Owners would simply be informed that their property had been listed, after which demolition or alteration could not be carried out without consent. Policies for preservation were, however, overshadowed by the need to reconstruct bomb-damaged areas and rehouse their former residents. The post-war dream promised to relocate people and employment in more healthy, rational and orderly patterns, with green belts to contain urban sprawl and new free-standing settlements inspired by Ebenezer Howard but instigated by the state through New Town corporations. There was broad support for the idea of a military-like onslaught on an unsatisfactory and outmoded urban environment to improve conditions for the citizens of a more just society.

All too often, however, post-war austerity and lack of vision seemed to produce disappointing, debased versions of Modernist ideals. By the mid-1950s, there were some who dissented from the very idea of a leisured, sanitised, motorised utopia and expressed nostalgia for the harsher but slower pace of pre-war Britain. Writing for *Punch* magazine in 1954, Betjeman described a visit to Middlesex, contrasting his observations with personal memories of the place some 40 years before: 'our buildings say to us: "We once were civilized. We are so no more."' Thus, he wrote his impressions of the new urban landscape along the 'Great Worst Road':

Warm air-conditioned world of beige and cream! Safe, labour-saving world of buff-tiled fireplaces, television sets and football pools! Hygienic world of community centres and culture, but not too much of it! There is no birth. There is no death. We will all go on for ever getting better and better. Break in, O bells of Brentford, from your fifteenth-century tower to remind us of the Truth.

(quoted in Denton 1988: 75)

Collins and Collins (1981) discuss the development of a British town-scape movement which emerged in architectural circles in the mid-1940s, and quickly gained momentum and popularity in the 1950s and 1960s. Its emphasis on incremental growth and democratic as opposed to tech-nocratic planning was in pronounced contrast to the prevailing 'masterplan' approach exemplified in the Comprehensive Development Areas, the favoured policy instrument for town centre redevelopment until the late 1960s. The word 'townscape' seems first to have appeared in the architectural press of the late 1940s, but the philosophy and outlook owed something to Pugin and Ruskin. There was also a revived interest in the work of Camillo Sitte, champion of the vernacular ideal in architecture and planning. First published in French and German in the 1880s, but only recently discovered by the English-speaking world, Sitte's influential book *Stadtebau* opens with an affirmation of the pleasures of visiting historic towns and cities:

> Enchanting recollections of travel form part of our most pleasant reveries. Magnificent town views, monuments and public squares, beautiful vistas all parade before our musing eye, and we savour again the delights of those sublime and graceful things in whose presence we were once so happy.
>
> (Sitte 1965: 3)

Sitte held engineers and technicians responsible for the failure of the modern city. Hope resided in enlightened and artistic architect-planners who might recreate what had developed anonymously and incrementally over centuries. 'Sitte was attracted by the innate sense of place and scale in plazas and streets that had been created by those who were actually using these spaces and were adapted by them over periods of time to fit the changing needs' (Collins and Collins 1981: 69). Unlike Pugin and Ruskin, however, Sitte recognised that 'modern living as well as modern building techniques no longer permit the faithful imitation of old town-scapes, a fact we cannot overlook without falling prey to barren fantasies' (Sitte 1965: 111).

Urban conservation and civic pride

Since the mid-twentieth century there has been a broadening appreciation of what might be worthy of conservation in the urban landscape. Buildings can be listed because of their 'special architectural or historic interest', but it has not been considered necessary to set out precise guidelines as to why a particular building should be included or excluded. It follows that a building is listed 'because expert opinion has decided that it has special character' (Pickard 1996: 1). It is important, therefore, to take account of the special-interest groups that emerged to proffer their expertise. Such groups are voluntary organisations set up to promote appreciation of particular historic architectural styles or themes, and to advocate protection where threatened by redevelopment, unsympathetic restoration or neglect. These organisations, in turn, have influenced public taste and public policy.

The Georgian Group (1937), originally a section of SPAB, was established to lobby for protection and to promote scholarly research relating to Georgian and Regency buildings. The society emphasised the integrity of complete terraces, crescents and squares. The Vernacular Architecture Group (1952) helped to promote awareness of traditional, regional building techniques and styles. The Victorian Society (1958) championed the then less-than-fashionable art and architecture of that period, including municipal, industrial and commercial buildings. The desirability of advancing statutory protection beyond the preservation of individual sites to the conservation of groups of building and urban spaces was advocated by the Civic Trust: an organisation set up in 1957 by Duncan Sandys MP to promote 'civic pride and good design' in the built environment. The Trust gained favourable media coverage and scored an early success in opposing John Cotton's redevelopment scheme for Piccadilly. The civic amenities movement gained a broad base of support, especially in the South of England, with 700 local societies affiliated to the Civic Trust by the mid-1960s (Ross 1991: 26).

The Civic Trust emphasised new uses for older buildings as well as imagination and respect for locality. The impact of architecture and planning that ignored these principles, served to strengthen their argument and support for local amenity groups. In the 1960s, startling structures using modern construction techniques housed new functions including airports, shopping complexes, blocks of open-plan offices, leisure centres and polytechnics. In addition, high rise flats inspired by Le Corbusier were developed and urban motorways swept into inner city areas. Casson (1975) reflects on the driving force for change in the 1960s and the emotional reaction it often provoked:

> 'the image of a white-hot, quick-moving technological sprint with money as the prize. The nation seemed swept off its feet by the speed of change, as we bobbed helplessly about in its wake, we searched in vain for the reassurance of familiar landmarks'. This new architecture of mass solutions seemed 'over-large, repetitive, shiny-faced and hard-edged ... to which we, the consumer, seemed powerless to contribute. Inevitably, public disenchantment with what they saw developed into public petulance'.
>
> (Casson 1975: 226–7)

A decade after the Civic Trust was founded, its chairman was able to instigate an important development in planning legislation which enshrined the concept of area-based conservation. Duncan Sandys, by then an opposition MP, was fortunate enough to win the necessary ballot to introduce a Private Member's Bill, with government support. Under the 1967 Civic Amenities Act, local authorities were henceforth required 'to determine . . . areas of special architectural or historic interest, the character of which it is desirable to preserve or enhance'. The designation of conservation areas under the Act was subsequently applied in a wide range of contexts.

The case of Covent Garden provides an illustration of the remarkable shift in public policy that occurred between the mid-1960s and the mid-1970s. The area had an established resident population and contained many historic buildings, with a range of small businesses, including craft activities, printers, costumiers and specialist retailers. Nevertheless, relocation of the fruit and vegetable market to a new site in Nine Elms, Vauxhall provided a unique opportunity for the Greater London Council (GLC) to replan a large area in the centre of London. Proposals published in the Comprehensive Development Area Plan (1968) included office blocks, high-rise housing, hotels, sports and conference facilities, as well as two new roads. Public reaction was overwhelmingly hostile and the plan became the subject an acrimonious debate between the GLC and an alliance of conservationists and the Covent Garden Community Association (Inwood 1998: 883–4).

Those who opposed the plan gathered wide support and favourable media coverage. Eventually, Lady Dartmouth, chair of the GLC's Covent Garden Committee, bowed to public opinion and resigned in 1972, issuing a frank statement: 'No individuals or bodies who represent the general public have supported us, and I have felt increasingly that our proposals are out of tune with public opinion with fears that the area will become a faceless concrete jungle.' In the following year, the Secretary of State for the Environment announced that, whereas he endorsed the proposed redevelopment, a further 245 properties would be listed. The scheme thus became impossible to implement (Richardson 1995: 278–9). The Covent Garden Action Area Plan (1978), in marked contrast to its predecessor, stressed the need to respect the area's unique character, especially through Conservation Area designation.

The desire to defend architectural heritage was expressed in another wave of conservation groups, a notable example being the Thirties Society, set up in 1979. There was also a National Piers Society (1980), a Railway Heritage Trust (1985), a Historic Farm Buildings Trust (1985) and a Fountains Society (1985). Hewison (1987: 24–6) commented sardonically on this trend, noting that the number of listed buildings was approaching half a million, and that well over 5,000 conservation areas had been designated. Not only had the pace of the conservation movement quickened, but the object of its concern had come closer to the present day. How long would it be before a Fifties Foundation or a Sixties Society? In fact, concern for architecture of these periods was subsequently espoused by the Twentieth Century Society, as the Thirties Society was renamed in 1992. Futhermore, a rolling '30 years rule', introduced by the then Department of the Environment, allowed any building of that age to be listed (Cullingworth and Nadin 1994: 161–2).

Commercialism, decadence and tourism

The prevailing mood of nostalgia was also to be exploited in new trends of contemporary building design. Postmodern architects of the 1980s combined an eclectic mix of historic references in pastiche: styles appropriate to the culture of private affluence, with scant regard for the squalor of public provision. 'Public space' was far less concerned with civic pride than with creating exclusive spaces conducive to hedonistic leisure consumption. The accent was on the design of individual buildings, rather than planning and this attitude to space was no longer shaped by the needs of the wider social project (Bianchini and Schwengel 1991: 214). Lord Rogers (1992) commented:

> Postmodern buildings are now decorated in pretty costumes: Neo-Classical, Gothic, Egyptian or Gipsy Vernacular. Post-Modernism, obsessed with money and fashion, has not produced a rigorous design or a better environment, for it cannot offer solutions to a world in need of an architecture that deals with the lack of public space, with the greening of the environment, with shelter for the less fortunate, with machines, flexibility and change.
>
> (R. Rogers 1992: 26)

In parallel with the exploitation of retro-styles in new buildings, a review of the historic built environment was to reveal the potential of the nation's heritage to generate revenue and earn its keep. In accordance with the philosophy of the Conservative government (1979–97), the 1983 National Heritage Act emphasised a market-led approach to the presentation and marketing of ancient monuments in state ownership or care. The Act established a new arm's-length agency: the Historic Buildings and Monuments Commission. Better known as English Heritage, it was to encourage greater public awareness of the sites and a more commercial approach. The attraction of 'industrial heritage' was demonstrated in centres such as Ironbridge, Beamish and Wigan. The potential of 'maritime quarters' was seen in Liverpool, Bristol, Swansea and other former dockland areas. A cluster of attractions could help an area project a clear and positive image of itself, as evidenced in Manchester's Castlefield Basin, where an area rich in industrial heritage has developed a thriving visitor economy (Law 1994: 78–82).

Some criticised the contrived use of historicised urban landscapes to market places as leisure and tourism destinations. Brett (1993), for example, refers to the Castle Museum in York, which in the early 1950s created a street and square of period shops and workplaces in the context of post-war reconstruction and modernisation. From the late 1970s:

> the toytown model ... spread outward into the town around the museum and come to bear down upon and to create a new civic reality. Central York is now dominated by the tourist economy and increasingly resembles its own museum; it is possible to walk through the sixteenth century alleys around the Minster and to think you are, indeed, within a museum, so antiquarian has the ambience become.
>
> (Brett 1993: 185–6)

Likewise Covent Garden, although saved from comprehensive redevelopment, underwent a fundamental change in character. Here, and in other schemes, regeneration involved an extensive rehabilitation of historic buildings and streets to present a clean, bright and cheerful face that would attract investors, visitors and new residents (Tiesdell, Oc and Heath 1996: 207–8). In some cases, refurbished heritage buildings could be hard to differentiate from their postmodern imitators. The overall effect of the new-old Covent Garden reminded York (1984: 46–7) of stage-craft such as the reconstructed shopfronts in museums such as the Museum of London or Disney World in Florida:

> If you look at the kinds of businesses that now predominate round there ... and the facilities – themed restaurants and wine bars which have *all been designed* – you realise that Covent Garden is the tip, the harbinger of the modern world, of real-life theme parks.
>
> (York 1984: 46–7)

Ashworth (1988) suggested that the construct of a 'historic city' was something more than the sum of its historic parts. Rather, it was created by the active intervention of institutions from the available stock of antiquities, using legal instruments and planning procedures in an attempt to conserve a particular interpretation of the past. At a deeper level, there was a 'conservational philosophy': an approach to urban governance which facilitated 'the functioning of the city for both residents and non-residents in such a way that the historical attributes are consciously exploited' (Ashworth 1988: 163). Although a symbiosis between urban conservation and tourism development was often assumed, Ashworth highlighted the possibility of tension between institutions whose accountability, practices and goals differed. This was manifest, for example, in the bowdlerisation of the historic city as a consequence of its attempts to meet the expectations of visitors who require a heritage experience that is simply, easily and quickly communicated. In general, few vistors have the time or inclination to accept the complications and subtleties of the city's historic development and the generally confused mix of time periods and architectural styles that typify most urban landscapes.

In many cases, promotion of tourism in historic centres seemed to suggest an escape culture which denied present-day reality (Goodey 1994a: 153–79). Some argued that this was symptomatic of a deeper malaise: the reactionary social attitudes of a post-imperial, post-industrial nation that had failed to come to terms with the reality of its position in the world. Lowenthal (1985: xvii) characterised English attitudes to the arts and the built environment as 'permeated by antiquarianism' with a marked bias 'in favour of the old and traditional, even if less useful or beautiful than the new'. This obsession with past glories was, perhaps, a sign of decadence or an abdication of responsibility for the present and future. An ascendant mood of the 1980s, nostalgia bolstered the ideology of the New Right and offered a discourse that obscured the conflicts and tensions of the past and denied the continuity of history into the real world of the present. It presented 'a unifying spectacle, the settling of all disputes. Like the guided tour as it proceeds from site to sanctioned site, the national past occurs in a dimension of its own' (Wright 1985: 69). New museums and heritage centres were opening at a phenomenal rate and an increasing proportion of the built environment was being conserved because of its historic interest (Hewison 1987).

Others have viewed the growth of a heritage industry in a more positive light. In an 'old country', a large stock of historic buildings may dominate the urban landscape. As functions change, redundant structures must be demolished, or else adapted to new uses that generate sufficient resources for conservation and subsequent upkeep. Leisure and tourism cannot provide a panacea for dereliction and economic decline. Nevertheless, they can facilitate a more secure future for historic buildings and contribute to a balanced strategy for urban regeneration. Through town trails, museums and heritage centres, presentation of a wide range of historical themes has become more imaginative and available to a wider audience. The experience need not be superficial if visitors are allowed to ask questions and make their own judgments. As Lumley (1994: 66) comments, Hewison's presentation of a monolithic heritage industry tends to characterise the public as a credulous mass, easily seduced by the sirens of nostalgia. Urry (1995: 145–6) refers to the aesthetic reflexivity of the viewer, stressing the ability of people to evaluate their society and its place within the world, both historically and geographically.

In this context, the distinctiveness of localities acquires greater significance. Shields (1991: 207–51) discusses place-myths of 'the British North – Land of the Working Class' perpetuated through literature, film and television. He emphasises the way in which the image has been appropriated and reworked in indigenous narratives, that have translated this mythical North into their own place-identity, and have cherished the images first propounded in the literature of (southern-based) writers since the nineteenth century. Robins (1991) notes a fascination with vernacular motifs, with reference to a cultural localism which reflects:

> deeper feelings about the inscription of human lives and identities in space and time. There is a growing interest in the embeddedness of life histories within the boundaries of place, and with the continuities of identity and community through local memory and heritage.
>
> (Robins 1991: 34)

The tourist's pursuit of authenticity (MacCannell 1976) may thus provide an opportunity for inner city areas and regions remote from the metropolis to break the cycle of disadvantage. It may encourage the rediscovery of long-neglected urban and industrial histories or a sense of local pride as advocated by the Civic Trust.

It cannot be assumed, however, that urban tourism will necessarily be informed by narratives told first hand by an 'indigenous' population. Goodey (1994b) comments on the development and marketing of historic quarters inspired by small-town and city core revitalisation programmes in the United States where partnerships between business, community and local authority were essential for success. He notes that in Britain the business–local authority linkages tend to be well established, but the wider community may be less involved:

> In the majority of cases residents are few (and getting fewer) and users, especially visitors, are seen as participating through the market mechanism alone. Concepts of participation and partnership are between the public authority and retail and development communities. As a result, use and manipulation of heritage tends to draw on an officially-recognised and/or marketable selection, abstracted from a more rooted community past.
>
> (Goodey 1994b: 20–1)

This recent expansion of leisure within, or associated with, spaces of retail and consumption activities raises issues concerning public leisure space and social exclusion as discussed in Chapter 2. Where commercial leisure provision and speciality retailing oriented towards visitors takes priority, some sections of society may be made to feel less than welcome and they may also form part of the very groups who rely most on the availability of public space to satisfy their social and/or leisure needs.

As Relph (1976: 55) emphasises, there is a fundamental distinction between the insider-participant and the external observer. The very idea of 'landscape' is associated with the gaze of the latter, since it implies a certain objectivisation of the scene and a distancing of subject from object. This way of seeing is also conveyed in the comparatively recent conceptualisation of 'townscapes'. In the context of historic quarters, townscape or streetscape improvements involve partnerships between particular stake-holders, especially individual property owners and businesses, public authorities and civic amenity societies. Such agencies must necessarily collaborate in the wider social project to achieve a harmonious effect. Many local residents and other users of public space may not, however, own land or buildings, run a business, or belong to a civic amenity society. Few would find it easy to use words such as townscape or streetscape to describe their everyday environment although their attachment to place may be profound: 'The composition of their landscape is much more inte-grated with the diurnal course of life's events – with birth, death, festival, tragedy – all the occurrences that lock together human time and place' (Cosgrove 1998: 19).

A conflict of ideals may therefore occur between urban conservation to maintain a sense of identity for local communities, and conservation that manipulates the urban landscape to satisfy visitors' expectations. Historic buildings are chance remains that seldom provide clues to an area's social or economic history for the casual observer. Tiesdell, Oc and Heath (1996: 211) argue that to make a historic quarter attractive to visitors, attention should be paid to the *permeability* of urban space: the ease by which people find their way into the area, as well as *legibility*: the facility to 'read' the character and history of the locality and to navigate their way around. Nevertheless, a number of important questions need to be addressed. If the urban landscape is to be selectively conserved and enhanced, who should provide the necessary expertise? If it is to be inter-preted, how to avoid the impression that one reading is unassailably correct? To what extent does a place-image created for external audiences correspond to the insiders' sense of place, if indeed there is a consensus among the latter?

The historic quarters of London's City Fringe

The City Fringe is an area of inner London adjacent to the Square Mile, one of the world's top three financial centres. It contains some of the cap-ital's least 'discovered' historic quarters: neighbourhoods that survived air raids as well as comprehensive development area schemes in the post-war era. Over many centuries, the identity of the Fringe has been shaped through its symbiotic, if unequal, relationship with the City. For many centuries, it has been regarded as a 'place on the margin' (Shields 1991) not only in location, but also in social status. From medieval times, the 'liberties' north and east of the precincts of the City provided space for excluded social groups, activities and institutions, many of which were prohibited or restricted within the city walls. Thus, it accommodated successive waves of migrants from other parts of the British Isles, as well as foreigners, many of whom sought refuge from religious or political persecution. From the fourteenth century, Flemish clothmakers were invited to establish their trade. From the sixteenth century, Sephardic Jews from Spain and Portugal set up businesses as merchants and moneylenders, while Protestant Huguenots from France brought their craft skills and business acumen (Porter 1996).

Its streets and open spaces provided the setting for some of London's most important fairs, festivals and markets and, a year after the proclamation of 1575 which banned plays within the City precincts, The Theatre opened in Shoreditch. This was England's first puropose-built playhouse on whose stage many of Shakespeare's plays were first performed (Cox 1994). Later, spas, pleasure gardens and other diversions attracted wealthy visitors to a relatively open landscape. From the early eighteenth century, however, its villages expanded rapidly, and much of the area became densely populated forming the squalid and overcrowded neighbourhoods that housed some of London's poorest inhabitants. Its public places now provided a platform for dissenters: nonconformists, trade unionists, reformists and revolutionaries. The spectacle of wealth and abject poverty juxtaposed was vividly described by novelists, including Defoe and Dickens, and portrayed in the engravings of artists such as Hogarth and Dore. The real-life horrors of the Jack the Ripper murders around Spitalfields and Whitechapel in the 1880s further reinforced the sinister side of the place-myth. Thus, the area became 'known' to many, evoking feelings of revulsion and fear as well as fascination.

By the early twentieth century, a wide range of manufacturing industries was located within the Fringe. Entrepreneurs made full use of the cheap labour of local residents, including thousands of recent Jewish immigrants from Eastern Europe. Factories and workshops, many with oppressive working conditions, developed cheek by jowl with densely packed housing. Particular trades became associated with particular areas, such as furniture, leather and clothing manufacturers in Shoreditch. Traces of these industries have survived to the present day, but the post-war era has seen a steep decline in manufacturing. Most of the Jewish population moved away in the 1960s and 1970s, but new immigrant groups took their place, adapting the old building stock for their own use and acquiring small businesses. Spitalfields is now home to a large Bangladeshi community, as well as a growing Somali population. With a diminishing manufacturing base, these ethno-racial groups have experienced high levels of unemployment. Their disadvantage has been compounded by poor housing stock and ageing infrastructure. Like those who lived in the area before them, they have also suffered harassment and periodic violence from race-hate organisations.

In recent times, however, there has been significant restructuring of the local economy in some parts of the Fringe. In the 1980s and 1990s, spectacular growth of financial services and other key sectors spread beyond the City and West End. Much of this expansion resulted in relocation to the 'new city' of London Docklands, but there has also been redevelopment at the borders of the Square Mile and the Fringe. Physical incursion into local communities and threats to familiar landmarks caused tension, a notable example being the proposed demolition and redevelopment of Spitalfields Market: a scheme successfully opposed by a local pressure group in the early 1990s (Woodward 1993). Some pockets of commercial and residential gentrification also emerged. As in other inner city neighbourhoods that have developed into 'urban villages' (Zukin 1982), the process was often initiated by artists and craftspeople acquiring studio and living space. In time, these areas attracted a cluster of design, public relations, media and other agencies servicing the City as well as wealthy residents. These, in turn, began to support a lively collection of bars, cafes and restaurants where the new service providers could entertain colleagues and clients: a nascent visitor economy.

In areas where Victorian industrial buildings and warehouses stood vacant, planning policies have encouraged refurbishment and conversion to new uses. These include studios for craft industries, as well as places for leisure and entertainment ranging from art exhibitions to live performances, recording studios and nightclubs. In Clerkenwell, within a conservation area and conveniently close to the City, a company called Manhattan Lofts began to convert older premises into fashionable apartments in the early 1990s. On the one hand, such investment generated new wealth, facilitated new uses for old buildings and led to the development of a complementary mix of activities. On the other hand, there was concern that established residents and firms would be displaced. Furthermore, the wave of new City-oriented businesses and residents locating in the Fringe had done little to enhance employment prospects among local communities. In the mid-1990s, 22 per cent of the resident workforce of around 30,000 people was registered as unemployed, with two-thirds out of work for six months or more. In general, local residents had not been able to participate in the boom. Fewer than 4 per cent of jobs in the Fringe, and only 1 per cent of City-based jobs were held by Fringe residents (City Fringe Partnership 1996: 5). As in other gentrified neighbourhoods, a white middle class with too much work has moved into an area with a high black and Asian population suffering from too little work (T. Butler 1996).

In order to tackle these problems strategically and across administrative boundaries and functions, the City Fringe Partnership (CFP) was set up with three other local authorities and other agencies in 1995 under the leadership of the Corporation of London. With funding from the government's Single Regeneration Budget (1996–9), a key feature of the programme of action was a 'Developing Cultural Quarters Project', the aim of which was to generate employment by encouraging visitors and marketing products. The project was designed to enhance the accessibility and appearance of key areas, including the historic quarters of Clerkenwell and Spitalfields/Brick Lane (City Fringe Partnership 1996). The Partnership highlighted the appeal of historical themes as well as its contemporary cultural scene for target markets including City businesspeople in addition to domestic and international visitors seeking alternatives to 'mainstream' tourism attractions.

A study was subsequently commissioned to consider the prospects for marketing the Fringe as a destination for leisure and tourism to these audiences. The consultants (Discover Islington and Mazorca 1997) emphasised the barriers that inhibited access to and movement within the newly defined cultural quarters. The City Fringe is a recent administrative construct and did not appear as a coherent entity in guidebooks used by visitors. Underground and surface rail services connect key nodal points to the City and elsewhere in the capital but the cultural quarters do not feature as station names on the tube map which acts as an important influence on people's orientation. Furthermore, account should be taken of its perceived accessibility from the City and other parts of London. The cultural quarters are within half an hour's walk of the City boundary but, to many Londoners and other visitors, the psychological distance is much greater. The best way to explore the area is on foot, but traffic noise and poor air quality are not conducive to a casual stroll and pedestrians must negotiate busy roads. Furthermore, the boarded up buildings and vacant sites, visible neglect, dirt and shabbiness create an unwelcoming impression and some visitors may feel uneasy, especially at night.

As the study also noted, few aspects of the area's eventful past are obvious from the buildings or streetscape. Although significant features of historic buildings and streetscape remain, much has been altered to accommodate new uses and, to the casual observer, there are few indications of previous occupants and activities. A notable example is the Neuve Eglise in Spitalfields. This is one of eleven Protestant churches built by Huguenots who were an expanding minority group during the early eighteenth century. The original congregation declined as the French-speaking minority became absorbed and, in 1809, the building became the London Society for Promoting Christianity among the Jews. Ironically, by the end of the nineteenth century it had become the Great Synagogue, used by Orthodox worshippers from Eastern Europe. As the local Jewish population declined in the 1960s it closed again, only to reopen in 1976 as the London Jamme Masjid: the great mosque for an expanding Muslim population (Hebbert 1998: 173). Such features of the streetscape may suggest the need for interpretation to reveal the area's rich history to onlookers, but there is also a need for sensitivity. In many cases, such as places of worship, the general public do not have right of access, and the host community may resent intrusion into their daily lives and culture.

At the time of writing, the development and marketing of the City Fringe and its cultural quarters is at an early stage, but some significant patterns and processes are emerging. In order to overcome the problems of accessibility highlighted above, programmes of action have prioritised improvements at 'gateways' and along the main thoroughfares. These have been complemented by better pedestrian signage to guide visitors from public transport terminals to attractions, as well as improvements to street lighting. In the case of Clerkenwell, a walking tour devised by a local history society has been turned into a permanent feature inspired by the Boston Trail in the United States. In 1998, the Culture Secretary opened the Clerkenwell Trail, a pedestrian route just over three kilometres long, marked out by plaques and banners designed by a local artist whose illustrations feature on an accompanying map. These and other local guides are made available through a small visitor centre staffed by volunteers. The tone of the Trail is discreet and unobtrusive, in sympathy with the medieval pattern of winding streets and alleys, as well as accommodating views expressed by local residents and businesses.

In the Spitalfields/Brick Lane cultural quarter, there has been a similar emphasis on accessibility, but the approach has been less restrained. With additional funding from Eastside City Challenge programme, the area will become a 'showpiece for London'. Again, first impressions are being enhanced, in this case by features such as gateway arches and other street furniture inspired by Asian motifs. Signage is also being improved, especially through Brick Lane and Petticoat Lane markets, and grants have enabled shops and restaurants to upgrade the streetscape (City Fringe Partnership 1997: 6). There are proposals to refurbish a number of visually important sites that are currently derelict or underused, including a listed market building which could house a bazaar/souk. There are also plans to expand existing attractions and create new ones, most notably the London Cultural Heritage Centre as a 'flagship' to 'foster a sense of pride amongst the local community and promote an image of London as an exciting and vibrant multicultural city. It will form the perfect complement to the traditional English history embodied by the Tower of London' (London Borough of Tower Hamlets 1996: 14).

The regeneration strategy for the City Fringe highlights issues that have wider significance for place marketing. There is a need to overcome negative stereotyped images and to make visitors aware of the rich heritage associated with the area's immigrant communities both past and present. The Cultural Quarters Project is envisaged as an asset to London as a whole and its prestige should raise the self-confidence of its communities. But the new positive images presented to outsiders must harmonise with the population's sense of identity and take account of their cultural diversity. There is a danger that complex truth may be simplified to create a strong storyline, and that the area may be turned into something of a folk museum, with historic buildings signposted as exhibits. Local people may be viewed as resident actors to animate the scene, their festivals and other customs presented as tourist attractions. The idea of the Fringe as a marginal place, may thus be reinforced by an 'exotic spectacle' and one-way traffic between the viewers and the viewed. Much will depend on the sensitivity with which the neighbourhoods are presented to visitors.

As Ashworth (1994) has argued, the raw materials of heritage – historical events, personalities, folk memories, mythologies, literary associations and surviving physical relics, together with the places with which they are symbolically associated – are converted into products through selection and packaging. Interpretation is literally the product.

Overview

To identify, present and interpret the 'valued legacy of the past' requires the appointment of experts to exercise the discrimination of taste. The nineteenth-century intellectuals who appointed themselves arbiters of this taste appealed to the sensibilities of a wider middle-class audience for their approval. Ancient relics and historic monuments were to be viewed with respect. As symbols of cultural and national identity, their physical presence reminded the onlooker of the country's past achievements. They deserved protection, not only from demolition, but from insensitive alteration that would destroy their integrity. Despite the opposition of landowning interests, the first steps had been taken towards statutory protection by the end of the nineteenth century. Voluntary action to acquire and maintain them had also begun. Sites and buildings thus preserved would provide an invaluable source of evidence to advance the study of archaeology and architectural history. Furthermore, they would serve an important role in public education. As with museums, art galleries and libraries, the legacy would be made available to contemporary audiences and preserved to inspire future generations.

The surveying and classification of historic buildings was initiated by voluntary societies, organisations that conducted scholarly research within their particular period of interest as well as campaigning for preservation. Through their advice to agencies of the state, such groups exerted a strong influence in defining the boundaries of inclusion. In the post-war era, protection of historic features of the built environment broadened considerably in scope with an increasingly liberal interpretation of what should be included. This 'democratisation' of heritage owes much to special interest and amenity groups, with passive support from a much larger audience (Lowe and Goyder 1983). The impetus seems to have come, in part, from reaction against architectural expressions of Modernism. This is especially so in relation to reaction against large-scale schemes to redevelop town centres, an issue that galvanised public opinion and profoundly influenced public policy. Since the mid-1960s, support has been gained for the principles of area-based conservation, incorporating good design and valuing features of the urban landscape that may have deep significance for people living, working and enjoying their leisure time in the locality.

The aesthetic enjoyment of heritage sites as visitor attractions is an important theme. If authenticity is subservient to visual appeal, it matters little to the onlooker whether the subject is an untouched antiquity enhanced by the patina of age, a thin facade or a replica (Larkham 1996). Most historic buildings have been renovated and altered many times, not only to accommodate new activities and functions, but also to satisfy changing tastes. Furthermore, Romantic sentiment has encouraged revivalism in the form of new buildings that make reference to historic styles. In some cases, it has also encouraged a rather cavalier approach to the renovation of genuine remains, conflicting with the purists' argument for preservation of relics in their original state, as well as their didactic role as evidence of the nation's cultural development. In recent times, whole streetscapes have been cleaned, renovated and represented to visitors as historic quarters. In the case of London's Covent Garden, a place that was threatened with large scale demolition, the rescued heritage now provides the setting for a commercially successful mix of cultural tourism, bars, restaurants and speciality retailing.

Such trends may be construed as part of a broader shift in leisure patterns and behaviour, where emphasis is placed on conspicuous consumption and pleasure, as demonstrated in shopping malls and theme parks. Consequently, the 'more traditional forms of high cultural consumption, such as museums and galleries, are revamped to cater for wider audiences through trading-in the canonical, auratic art and educative-formative pretensions for an emphasis upon the spectacular, the popular, the pleasurable and the immediately accessible' (Featherstone 1991: 96–7). Mindful of these developments in demand, and faced with rapid decline of manufacturing, economic strategies for many older industrial centres tended to focus on land and property, often through the creation of self-contained enclaves of leisure and tourism-related consumption (Bianchini and Schwengel 1991). In this context, full recognition was given to the image-value of historic urban landscapes in a competitive global market for inward investment, high-spending visitors and wealthy new residents. But, in the process of creating a favourable external gloss, the identity and interests of other stakeholders, including less powerful and less articulate local residents, small businesses and organisations, could easily be eclipsed.

In deference to the new millennium, some argue that morbid obsession with the past is best cured by destruction and reconstruction. There is, however, a more general desire to develop innovative ways of managing change in the urban landscape. According to Rossi's (1982) thesis, the built environment should link past and present, offering scope for future development in a pleasing mixture of forms and styles while avoiding the tyranny of a single present. The property and image-led strategies, characteristic of the 1980s and early 1990s, have been discredited and public policy now emphasises integration of conservation with the social objectives of regeneration. English Heritage (1998: 23) for example, will contribute to initiatives 'which by being grounded in existing character and meaning, and by involving a wide range of partners, including the communities themselves, can begin to address the problems identified by the Social Exclusion Unit'. In the case of London's City Fringe, implementation of this policy is at a formative stage and many uncertainties remain. There is, however, a strong expectation that its initiatives will improve the quality of life for local communities as well as the quality of experience for visitors through celebration of a pluralistic heritage that has hitherto been hidden or denied.

9   Landscapes of desire
Reappropriating the city

Introduction

In their introductory chapter to *Mapping Desire: Geographies of Sexualities*, Bell and Valentine (1995: 1) stated that one intention of their book was that it 'should at least provide an introduction to ways in which the spaces of sex and the sexes of space are being mapped out across the contemporary social and cultural terrain'. This chapter seeks to develop that intention in relation to social and cultural geographies of leisure and tourism. The chapter will begin by summarising some of the discussions of 'spaces of sex' and 'sexes of space' that took place in relation to geographies of leisure and tourism during the 1990s (Aitchison and Reeves 1998; Bell and Binnie 1998; Bell and Valentine 1995; Gluckman and Reed 1997; Knopp 1995, 1997; Lauria and Knopp 1985; Mort, 1996; Munt 1995; Valentine 1993a, 1993b, 1993c).

Many of these geographies demonstrate the 'normalisation' of heterosexuality within the landscape. Here, dominant and repeated performances of sexuality inscribe and repeat particular forms of spatial identity from a corporeal scale right through to the scale of the city itself. But, the contestation, subversion and transgression of such dominant spatial identities and patterns have become an increasing focus of contemporary social and cultural geographies (Aitchison 1999a, 2000b Skeggs 1999; Valentine 1996). Following a contextualisation of the sexuality of leisure spaces and places, this chapter seeks to develop discussion of three particular types of urban space that have been central to the contestation of heterosexual domination of leisure and tourism landscapes. The three landscapes differ in scope and scale from the city to the street and the hotel.

Our understanding of the duplicitous and often ambiguous identity of urban 'gay quarters' can thus be developed by situating these areas within the wider urban landscape while focusing in on the urban street as a micro landscape of that larger heterotopia so frequently referred to as the 'gay community'. Narrowing our gaze still further, an exploration of the environment of the hotel as a tourism landscape offers insight into the ways in which space is negotiated, contested, transgressed and appropriated in relation to sexuality.

Chapter 7 introduced gender as a key signifier in the representation and consumption of these leisure and tourism landscapes. Drawing on the concept of the gaze, outlined in Chapter 2 and revealed further in Chapter 5, notions of inclusion and exclusion, together with possibilities for disrupting and transgressing dominant definitions of space and place, were introduced. But, whereas Chapter 7 was primarily concerned with the place of gender in informing the representation and consumption of landscape, this chapter focuses on sexuality as a key spatial signifier in the production and consumption of leisure and tourism. Increasing attention has been given to the intersection of sexuality, space and identity in recent geographical literature (Adler and Brenner 1992; Bell and Binnie 1998; Myslik 1996; Valentine 1993a, 1993b, 1993c). Indeed, Bell *et al.* (1994) state:

> Work on what we might call lesbian and gay spaces has made equally apparent this notion of a mutual relationship between space and identity. It has been shown that sexual identity impacts on the use and reading of space, and that the socially and culturally encoded character of space has bearing on the assuming and acting out of sexual identities.
>
> (Bell *et al.* 1994: 32)

Queer space: material and symbolic landscapes

The previous chapter identified the interrelation between social, cultural and economic influences in the construction of cityscapes of modernity and nostalgia. So too have economic, social and cultural factors been interwoven in the development and display of gay and lesbian leisure. The discretionary spending power of gay men, together with the increasing recognition of the value of the pink pound by leisure and tourism providers, appears to have facilitated the increasing visibility of what might be termed 'landscapes of lavender leisure'. Thus, Binnie (1995: 182) has argued for a 're-incorporation of the *material* into emerging debates on sexuality and space'. Similarly, Gluckman and Reed (1997) have offered economic analyses of a range of social and cultural experiences of gay men and lesbians. Badgett (1997), however, in an attempt to overcome some of the economic stereotyping of gay men, has urged caution in assuming that gay men have higher disposable income levels than their heterosexual counterparts. In other words, while the increasing commodification, commercialisation and marketability of gay leisure lifestyles in the 1980s was undoubtedly linked to the economic boom experienced by young urban professionals, the homogenisation of the category 'gay men' presents a simplified and depoliticised perspective on a complex and heterogeneous group.

Mort (1996: 166), in his discussion of the increasing centrality of gay consumersism in 1980s' London, highlighted the assumed tension between conspicuous consumption and political activism. Instead of reducing political presence and pressure, however, Mort argues that increasing levels of leisure spending and consumption by gay men resulted in a 'consumerist ethos ... encouraging homosexual men to stake a greater claim to ownership of the city'. In the 1980s and 1990s this 'greater claim to ownership of the city' was clearly concerned with financial *and* cultural ownership and claims for space. In contrast to previous claims, however, the developments of the 1980s and 1990s related as much to public space as to the private, invisible or underground spaces of the established gay scene. Through the appropriation of commercial leisure and retail space in streets like Old Compton Street in London's Soho, the gay *scene* became *seen*, rendered visible by the gay community in an otherwise dominant heterosexual landscape. Rothenberg (1995), however, rightly summarises some of the difficulties in using the term 'community' within discussions such as this. She states:

> 'Community' is a tricky word, however, with numerous meanings and uses. Urban geography and sociology tend to use 'community' to indicate a geographically bounded area consisting of people who share particular characteristics, such as race, ethnicity and class, and who maintain social interactions with each other (Karp *et al.* 1991). In such discussions, 'community' is often used interchangeably with 'neighbourhood' (Johnston *et al.* 1986).
>
> (Rothenberg 1995: 171)

Parallel studies of the emergence of gay communities, neighbourhoods, districts or quarters have been undertaken in relation to a number of other cities, for example Amsterdam (Binnie 1995; H. Hughes 1997; Pritchard *et al.* 1998), Boston (Davis 1995), Brighton (Shields 1991), Manchester (H. Hughes 1997; Skeggs 1999) and Minneapolis (Davis 1987).

Gay destinations: the landscape of the city

Cities, like the streets within them, have been associated with ambiguity, anonymity and multiplicity of identities. Wilson states that:

> Sexuality, was only one source of threatening ambiguity and disorder in the city. The industrial city became a crucible of intense and unnerving contrasts ... Escape and entrapment, success and disaster offered heightened, exaggerated scenarios of personal triumph or loss of identity.
>
> (Wilson 1991: 5)

Wilson goes on to examine these contradictions or dualities in relation to the sexual duality of male–female. She sees the city as embracing the stereotypical dualisms discussed in relation to the gendered spaces outlined in Chapter 7. In this way:

> the city, a place of growing threat and paranoia to men, might be a place of liberation for women. The city offers women freedom. After all, the city normalises the carnivalesque elements of life ... despite its bureaucratic conformity, at every turn the city dweller is also offered the opposite – pleasure, deviation, disruption ... We might even go so far as to claim that urban life is actually based on this perpetual struggle between rigid, routinised order and pleasurable anarchy, the male-female dichotomy.
>
> (Wilson 1991: 5–6)

H. Hughes (1997) emphasises the significance of what he terms 'gay clusters' within cities. The complexity of urban centres is such that there may be more than one gay quarter or zone within the urban locale. For example, Hughes highlights four areas of Amsterdam which house gay leisure spaces: Waroesstraat, Amstel, Reguliersdwarsstraat and Kerkstraat. H. Hughes (1997: 4) goes on to state that 'each has its own distinct character, performing a different function and being frequented by a different "clientele". These spaces are leisure spaces and are on the margins of the "conventional" centres for entertainment'. In London, too there are clusters of gay leisure spaces and Mort comments that:

> Place, setting and context were crucial to the enactment of commercial forms of masculinity. A spatial matrix – of professional and business networks, of knowledges and representations – marked out a set of masculine personas which were associated with London as a consumer city. Yet it is here that the map needs rendering more precise. It was not London as a whole which was understood to be implicated in these spatial transformations. The urban geography was much more concrete. It was specific zones of the city, definite quarters, which became closely associated with these configurations of gendered commerce. Prominent among the areas was Soho, in the West End.
>
> (Mort 1996: 150)

Although to outsiders, the increasing visibility of gay space within London's Soho appeared to signify a fairly rapid transformation, the appropriation of leisure and retail space by the gay community had its origins in a much earlier era. In the late nineteenth century the sexuality of Soho's space began to be mapped out as a social and cultural community in addition to being a physical neighbourhood. Mort (1996) emphasises that:

From the late nineteenth century a plethora of guides and literary souvenirs emphasised that Soho did not possess an 'official' geography. Rather, its space could only be known by initiated tourists. Nevertheless, in terms of physical mapping the name was usually taken to designate the square mile of densely packed streets crossed by four major thoroughfares: Oxford Street to the north, Coventry Street to the south, Charing Cross Road to the east and Regent Street to the west. These boundaries functioned not only as physical extremities; they also carried strong symbolic resonances about the limits of Soho's cultural influence. Each of the main streets signalled a spatial resumption of more mainstream and normalising transactions.

(Mort 1996: 151–2)

Sexuality and spectacle: the landscape of the street

Previous sections of this chapter and the last have referred to the duplicitous, ambiguous and contested nature of streetscapes as sites of leisure and tourism. Just as such sites can be appropriated by new and emergent groups of previously marginalised tourists and leisure participants, so too can these landscapes be reappropriated by previously dominant users. Pritchard *et al.* (1998) highlight the ways in which gay spaces, sites and festivals are increasingly being reappropriated by straight tourists and visitors. In highlighting the 'significant opportunities for tourism marketers to reach out to the gay consumer' there is always a possibility

> that in doing so they may 'degay' spaces and events, erasing their essentially gay identities. Such developments run the risk of undermining the 'sanctity' of gay spaces and places thus threatening these hard won oases in a largely heterosexual world; a threat which has implications for gay people and their place in heterosexual societies.
>
> (Pritchard *et al.* 1998: 33)

But this so-called 'degaying' of leisure space has come at just the time when leisure and tourism providers are beginning to recognise the value of the gay market. Holcomb and Luongo (1996) have provided evidence of the economic value of the gay travel market and Pritchard *et al.* (1998) have referred to gay tourists as the closet thing to a recession-proof market. Identification of this lucrative tourist market has resulted in numerous gay travel companies such as Pride Travel and Touch Holidays, and the gay press is increasingly taken up with advertising by such companies and reports of travel to destinations identified as gay and gay friendly: Amsterdam, Florida, Gran Canaria, Ibiza, Kos and Sitges. In addition to these holiday destinations, a number of festivals celebrating gay pride have also become tourist attractions for gay travellers and spectacles for straight tourists: Sydney Mardi Gras, the New Zealand Heroes Parade, the

Manchester Mardi Gras, Edinburgh's Tea in the Park, and the Pride Festival in London. Pritchard *et al.* (1998: 44) call for those involved in tourism marketing 'to recognise the tensions which may result from their activities: tensions which may affect any events once they become commercialised and once communities become the target of mainstream marketers'.

Sexuality and hospitality: the landscape of the hotel

As Chapter 7 demonstrated, many policy-led attempts to increase women's participation in leisure failed to recognise women as a heterogeneous group. Moreover, previous feminist research within leisure studies makes few references to lesbians' leisure (Aitchison 1997; Hargreaves 1994: 260). Similarly, in tourism and hospitality studies there are few references to difference and diversity among women and the hotel is frequently identified as a site of heterosexual display. As Valentine has stated:

> In particular, hotels have a dual image; first they represent a surrogate home for families on holiday and therefore are associated with heterosexual family units; second, they are effectively surrogate bedrooms having specific (hetero)sexual associations as a site for adultery and 'dirty weekends'.
>
> (Valentine 1993a: 402)

Within the tourism and hospitality sector, specific provision for women has extended only to a limited number of pubs and clubs catering for a small lesbian market and to specialist activity holidays catering for women-only groups. Until recently, women-only provision did not extend into the commercial sector market for tourist accommodation. Over the last ten years, however, it has become possible to identify three distinct types of tourist space in the form of accommodation marketed specifically at women: tailored accommodation for business women, accommodation marketed at lesbians and offered within mixed-sex gay hotels, and women-only accommodation.

Although women business travellers represent a growing market segment in the hotel industry, market analysis of the relationship between gender and business travel is limited. This lack of empirical research, however, has not deterred some of the major hotel chains from developing initiatives to capture this market. For example, Trust House Forte, the Holiday Inn group and Thistle Hotels have all implemented policies and practices to attract women business travellers. Their efforts have been addressed at increasing the standard of provision in general and improving security measures in particular. Following evidence from women guests, combined with some limited research by the hotels themselves, there are indications that such approaches may be interpreted as paternalistic or even patronising by women, and a number of these 'products' have now been withdrawn.

Trust House Forte were first to recognise the importance of the woman business traveller market when their chain of Forte Crest hotels launched Lady Crest in 1983. Holiday Inn then offered the Lady Business Guest service at their six Crown Plaza Holiday Inn hotels from 1991. More recently, Thistle Hotels introduced a policy to make women travellers feel especially welcome. Trust House Forte's Lady Crest service was discontinued following negative feedback from women guests. Research by the hotels themselves indicated that women sometimes felt more visible and vulnerable by being singled out for special treatment. For example, the Lady Business Guest service offered by the Holiday Inn focused upon security measures to minimise the threat of harassment by male guests and hotel workers. In doing so, however, eight out of ten 'promises' explained in the Lady Business Guest leaflet served to remind women of their vulnerability: 'with your safety in mind, a lady guest's car is parked in a space located as near to the hotel as possible and an escort to the car is available'. While such chaperoning and surveillance techniques recognise the threats to women travelling alone, they do not address the wider issues of women's fear of public space. Indeed, such measures could be seen as a rather ironic juxtaposition when placed against other hotel services such as the availability of 24-hour pornography channels.

The second type of distinct tourist space for women, identified above, is that of accommodation marketed at lesbians and offered within mixed-sex gay hotels. Previous research has indicated, however, that women are not treated as equal to men in the provision of mixed gay and lesbian accommodation (Aitchison and Reeves, 1998). This type of accommodation is aimed at men, and although women are tolerated in some establishments their presence does not appear to be encouraged. Although there has been increasing talk of the value of the 'pink pound' in recent years, it would seem that the economic power of the pink pound is invested in male rather than female hands in relation to both providers and participants in tourism.

Interviews with the managers of three of London's most popular gay hotels confirmed that women were tolerated rather than targeted as guests. One hotel which prides itself in being 'London's largest and busiest gay hotel' estimated that only 5 per cent of guests were women and that the hotel attracted 'girls who want to have a laugh and a good time, not those politically correct feminists'. The same manager stated, however, that he did not think he received any repeat business from women and that he was not really interested in why this might be so. The only female member of staff employed in the hotel was a cleaner and there were no lesbian staff. The manager stated that men expected and preferred male staff and that he did not know what women expected or preferred. This hotel had designated the top floor as men only and, when asked if he thought it might be appropriate to have a women-only floor, the manager said that he thought not, as the women liked to be among the men. The manager then went on to say that he would prefer to accommodate any women in an annex away from the main hotel thus increasing the spatial demarcation between gay and lesbian guests.

While the hotel discussed above made no efforts to attract women guests, other hotels actively discouraged women. For example, one hotel that is featured in the *Travel Her Way* directory and described as being 'operated by two gay men who welcome lesbians and gay men from all over the world', features only men and male couples in their publicity material.

The third type of accommodation cited above is that of women-only accommodation. In a recently published directory, Bonner and Isles (1995) list a total of fifty-five UK guest houses and bed and breakfast establishments for women. However, only thirty-six of these are listed as 'women-only' with the remainder being described as either 'women run' or 'women friendly'. The largest women-only provider of tourist accommodation in the UK was Reeves Hotel which operated in London between 1988 and 1997.

The rationale behind the development of Reeves Hotel was to provide accommodation space where women could be guaranteed good service and an absence of any unwanted harassment. These factors were particularly important for women travelling alone on business who had reported unequal service, unwanted attention, and even harassment and abuse from male guests. The importance of women's safety, both inside and outside the hotel, was a high priority and the chosen location had to combine affordability with safety considerations such as good public transport links, off-street parking, well-lit access and some privacy in the form of a secluded garden. The eventual site, a Georgian terrace overlooking Shepherd's Bush Green in West London, was converted from its previous use as office accommodation. Access to public transport was an important factor in choosing the location and the hotel was a two-minute walk from the nearest underground station and had excellent transport links to central London and Heathrow airport. In addition, the property had space for off-street parking and there was an overgrown garden that could be converted into a secluded patio garden. The economic success of the hotel was mixed and the most severe setback to the business took place within two years of opening, when interest rates in the UK doubled from 7 to 15 per cent. At the same time, the hotel's main target market of businesswomen was also suffering from the impending economic recession as many companies cut back on business travel. Although the market of leisure travellers, which incorporated a large number of lesbians, remained fairly constant, self-funded guests tended to opt for cheaper accommodation and did not spend in the hotel's bar or restaurant.

When asked to rank the hotel's services and facilities in order of importance, both social and cultural factors related to the material and symbolic nature of leisure and tourism space were highlighted. The women-only environment was graded as essential by the largest number of respondents and the following factors were listed in descending order of priority: lesbian friendly, safety, proximity to public transport, price, location, private parking and private garden. The identification of the women-only environment as the most important factor in attracting women to the hotel was encouraging given the nature of the hotel, the rationale for establishing it, and the emphasis of the hotel's marketing. The importance of public transport in influencing women's leisure patterns has been documented in previous feminist leisure research (Deem 1986; Green, Hebron and Woodward 1987; Wimbush 1986). Similarly, safety issues and fear of violence have been seen as important in determining women's use of public space (Valentine 1989). Although the hotel was not established as a 'lesbian hotel', this group of guests formed a larger market segment in the later years than it did at the outset. This may be explained by the shift from a predominantly business travel market to the leisure market combined with attempts by mainstream hotels to attract the women business travel market.

Although a socialist feminist analysis would emphasise price as being a major determinant of travel, this factor was not ranked as being one of the most important. In relation to women and travel, however, it is not a question of price differentials of a few pounds that are important, but the ability to afford to travel at all. For the majority of women in the UK, travelling to and staying in any London hotel is likely to be beyond their financial means. Within hospitality management literature, location is considered to be the most important factor in determining people's choice of hotels (Davis and Stone 1985). This body of literature, however, mirrors much of the positivist geography literature outlined in Chapter 3. The history of Reeves Hotel serves to demonstrate that spatiality, and the interrelation between spatiality and sexuality, add complexity to the seemingly straightforward world of tourist accommodation provision. Chapter 2 attempted to demonstrate the significance of the social and symbolic nature of place in addition to the physical and absolute nature of space. This chapter, too, has indicated that people can be 'in' and 'out' of place at different times and in different locations. Space is therefore an important constituent in 'the intersection of "sex", "gender", "identity" and space; with their construction and performance, their *constructedness* and their *performativity*' (Bell *et al.* 1994: 31).

10 Relocating landscapes
Leisure, tourism and culture

Introduction

The final chapter to a book is usually titled 'Conclusion' in reference to the provision of a fairly straightforward summary of the research and an attempt to offer 'closure' to the project. As our introductory chapter highlighted, however, our journey through constantly evolving leisure and tourism landscapes can never be fully complete. This final chapter is therefore presented more as a sojournment in our travels than as a definitive end to our travelogue.

The limitations of providing a full and final conclusion to a largely poststructuralist project are therefore acknowledged. Instead, this chapter seeks to present a less definitive end to our travels by engaging in four distinct but interrelated strands of discussion. First, the chapter will present a brief summary of each of the eight main chapters of the book. Second, the chapter will identify and comment on a range of themes that have emerged from these chapters. Concomitant with a poststructuralist approach, these themes all serve to identify and challenge dominant dualistic readings of landscape. Third, this chapter seeks to discuss the possibilities for further leisure and tourism geographies and geographies of leisure and tourism. Finally, the chapter asks that these future leisure and tourism studies find a place for 'culture' in their engagement with space and place.

Following the book's Introduction, Chapter 2 sought to situate social and cultural geographies within an overarching chronology of geographical theories that have engaged with leisure and tourism landscapes. This was the first of four chapters that demonstrated different 'ways of seeing' landscapes of leisure and tourism. Chapter 3 then illustrated ways in which our views of these landscapes have changed over time and have been shaped by developments in travel, transport and communications. Chapter 4 then sought to demonstrate that not only our *views* but the *values* we place on landscapes have changed over time. Chapter 5 then illustrated that these views and values have, to a large extent, been informed by different media representations of landscape and by different trends and patterns of consumption of these media.

In the second half of the book we moved from looking at different ways of representing or seeing landscapes to analysing different ways of consuming or '(re)placing' landscapes. These chapters examined heritage landscapes, gendered landscapes, urban landscapes and landscapes of desire, and illustrated that leisure and tourism landscapes can be differently constructed, consumed, interpreted and appropriated by different groups and individuals, in different ways and in different times and spaces. In other words, these landscapes of leisure and tourism, together with those who engage with these landscapes, are combined in a constant state of negotiation, renegotiation and transition. Chapter 6 demonstrated that the relationships between history and heritage and between past and present are blurred by the sites and processes of leisure and tourism. The case studies of heritage attractions in the south-west of England illustrated that landscapes can be 'multi-vocal', 'evolving' and 'mythical' rather than static representations of one reality or truth. Chapter 7 examined the evolving discourses of spatialised feminism and gendered space in relation to leisure and tourism landscapes. Here, the case study of one Scottish heritage development offered further illustration of the ways in which a particular heritage landscape has been represented in relation to masculinism, militarism and nationalism. Chapter 8 then offered an analysis of the ways in which space, place and landscape can be transformed over time to accommodate different social, cultural, economic and political priorities. The case study of London's City Fringe illustrated the constantly evolving nature of landscape and the constantly changing interactions which people have with the landscape around them. Chapter 9 then developed a further case study of London in its analysis of queer space and gay and lesbian leisure sites at the different scales of the city, the street and the hotel.

In all of these chapters we have been conscious of the recurrence of a number of themes. Together, these themes challenge the dualistic nature of many of our previous readings of leisure and tourism landscapes. Many of the chapters have rendered visible the false dichotomy or dualism between the self and Other, production and consumption, and inclusion and exclusion. Chapters 6 and 7, for example, with their focus on heritage, have illustrated the blurred boundaries between history and heritage, fact and fiction, past and present, and myth and reality. These discussions have emphasised the importance of what de Certeau (1984) stressed as the productive consumption of everyday life such that people and places are engaged in constantly evolving processes of interaction and 'remaking' of the world around them.

Although recent social and cultural geographies have engaged with de Certeau and others, the same is less true of leisure and tourism studies, where structuralist interpretations have remained dominant. Moreover, while these geographies have shed light upon the interrelation between leisure, tourism and spatiality, these studies have rarely been framed explicitly as leisure or tourism studies. This book has sought to develop *leisure and tourism geographies* to a point where they might meet *geographies of leisure and tourism*. In other words, through further reinterpretations of leisure and tourism landscapes scholars of leisure and tourism studies might become equally aware of social and cultural geographies and social and cultural geographers might become mutually aware of leisure and tourism studies. The increasing awareness of the place of culture and the contribution of cultural studies to both geographies and leisure and tourism studies may assist in this process. Poststructural theory, with its emphasis on the interrelation of structure and culture, and the material and the symbolic, remains central to this discursive development. Our aim, then has been to *relocate* landscape in relation to both geographies and leisure and tourism studies.

References

Abercrombie, P. (1938) 'Country planning', in C. Williams-Ellis (ed.) *England and the Beast*, London: Readers' Union and J. Dent & Sons.

Adkins, L. (1998) 'Feminist theory and economic change', in S. Jackson and J. Jones (eds) *Contemporary Feminist Theories*, Edinburgh: Edinburgh University Press.

Adler, S. and Brenner, J. (1992) 'Gender and space: lesbians and gay men in the city', *International Journal of Urban and Regional Research* 16: 24–34.

Adorno, T. (1993) *Dialetics*, London: Routledge.

Agyeman, J. and Spooner, R. (1997) 'Ethnicity and the rural environment', in P. Cloke and J. Little (eds) *Contested Countryside Cultures: Otherness, Marginalisation and Rurality*, London: Routledge.

Aitchison, C. (1997) 'A decade of Compulsory Competitive Tendering (CCT) in UK sport and leisure services: some feminist reflections', *Leisure Studies* 16, 2: 85–105.

—— (1999a) 'New cultural geographies: the spatiality of leisure, gender and sexuality', *Leisure Studies* 18, 1: 19–39.

—— (1999b) 'Heritage and nationalism: gender and the performance of power', in D. Crouch (ed.) *Leisure Practices and Geographical Knowledge*, London: Routledge.

—— (2000a) *Disability and Social Inclusion: leisure, sport and culture in the lives of young disabled people*, Cheltenham: Cheltenham and Gloucester College of Higher Education and Scope.

—— (2000b) 'Leisure and urban exclusion: developing leisure geographies and geographies of leisure', *The North West Geographer* 3, 2: 13–20.

—— (2000c) 'Poststructural feminist theories of representing others: a response to the "crisis" in leisure studies discourse', *Leisure studies* 19, 3: 127–44.

—— (2000d) 'Women in leisure services: managing the social-cultural nexus of gender equity', *Managing Leisure: An International Journal* 6, 4: 221–36.

—— (2000e) 'Locating gender: space, place and heritage tourism', in K. Atkinson, S. Oerton and G. Plain (eds) *Feminisms on Edge: Politics, Discourses and National Identities*, Cardiff: Cardiff Academic Press.

—— Brackenridge, C. and Jordan, F. (1999) *Gender equity in leisure management*, Reading: Institute of Leisure and Amenity Management.

—— and Jordan, F. (eds) (1998) *Gender, Space and Identity: Leisure, Culture and Commerce*, Eastbourne: Leisure Studies Association.

——, —— and Brackenridge, C. (1999) 'Women in leisure management: a survey of gender equity', *Women in Management Review* 14, 4: 121–7.

—— and Reeves, C. (1998) 'Gendered (bed)spaces: the culture and commerce of women only tourism', in C. Aitchison and F. Jordan (eds) *Gender, Space and Identity: Leisure, Culture and Commerce*, Eastbourne: Leisure Studies Association.

Alfrey, J. and Putnam, T. (1992) *The Industrial Heritage: Managing Resources and Uses*, London: Routledge.

Alonso, W. (1960) 'A theory of the urban land market', *Papers and Proceedings of the Regional Science Association* 6: 149–58.

Althusser, L. (1971) *Lenin and Philosophy,* London: New Left Books.

Alvesson, M. and Billing, Y. D. (1997) *Understanding Gender and Organizations*, London: Sage.

Andrews, M. (1989) *The Search for the Picturesque: Landscape Aesthetics and Tourism in Britain, 1760–1800*, Aldershot: Scolar.

Appleton, J. (1996) *The Experience of Landscape*, Chichester: Wiley.

Arnold, J., Davies, K. and Ditchfield, S. (eds) (1998) *History and Heritage: Consuming the Past in Contemporary Culture*, Shaftesbury: Donhead.

Ashworth, G. (1988) 'Marketing the historic city for tourism', in B. Goodall and G. Ashworth (eds) *Marketing in the Tourism Industry: The Promotion of Destination Regions*, London: Routledge.

—— (1994) 'From history to heritage – from heritage to identity: in search of concepts and models', in G. J. Ashworth and P. J. Larkham (eds) *Building A New Heritage: Tourism, Culture and Identity in the New Europe*, London: Routledge.

—— (1996) 'Elements of planning and managing heritage sites', in W. Nuryanti (ed.) *Tourism and Heritage Management*, Yogyakarta: Gadjah Mada University Press.

—— and Voogt, H. (1994) 'Marketing and place promotion', in J. Gold, and S. Ward (eds), *Place Promotion: The Use of Publicity and Marketing to Sell Towns and Regions*, Chichester: Wiley.

Badgett, M. V. L. (1997) 'Beyond biased samples: challenging the myths on the economic status of lesbians and gay men', in A. Gluckman and B. Reed (eds) *Homo Economics: Capitalism, Community and Lesbian and Gay Life*, London: Routledge.

Bagwell, P. S. (1974) *The Transport Revolution from 1770*, London: Batsford.

Bale, J. and Philo, C. (eds) (1998) *Body Cultures: Essays on Sport, Space and Identity by Henning Eichberg*, London: Routledge.

Barbier, B. (1984) 'Geography of tourism and leisure', *Geojournal* 9, 1: 5–10.

Barnes, T. and Gregory, D. (1997) *Reading Human Geography: The Poetics and Politics of Inquiry*, London: Arnold.

Barrell, J. (1972) *The Idea of Landscape and the Sense of Place, 1730–1840: An Approach to the Poetry of John Clare*, Cambridge: Cambridge University Press.

—— (1982) *The Dark Side of Landscape: The Rural Poor in English Painting 1730–1840*, Cambridge: Cambridge University Press.

Barthes, R. (1982) *Empire of Signs*, trans. R. Howard, New York: Hill & Wang.

Bell, David (1991) 'Insignificant others: lesbian and gay geographies', *Area* 23: 323–9.

—— and Binnie, J. (1998) 'Theatres of cruelty, rivers of desire: the erotics of the street', in N. R. Fyfe (ed.) *Images of the Street: Planning, Identity and Control in Public Space*, London: Routledge.

—— and Valentine, G. (eds) (1995) *Mapping Desire: Geographies of Sexualities*, London: Routledge.

—— and —— (1997) *Consuming Places: We Are Where We Eat*, London: Routledge.

——, Binnie, J., Cream, J. and Valentine, G. (1994) 'All hyped up and no place to go', *Gender, Place and Culture* 1, 1: 31–47.

Bell, Desmond (1993) 'Framing nature: first steps into the wilderness for a sociology of landscape', *Irish Journal of Sociology* 3: 1–22.

Belloc, H. (1942) 'The crooked streets', in W. W. Williams (ed.) *A Book of English Essays*, Harmondsworth: Penguin.

Bender, B. (1993a) 'Stonehenge – contested landscapes (medieval to present-day)', in B. Bender (ed.) *Landscape: Politics and Perspectives*, Oxford: Berg.

—— (1993b) *Landscape: Politics and Perspectives*, Oxford: Berg.

—— (1998) *Stonehenge: Making Space*, in B. Bender (ed.) *Landscape: Politics and Perspectives*, Oxford: Berg.

—— and Edmonds, M. (1992) 'Stonehenge: whose past? What past?', *Tourism Management* 13: 355–7.

Berger, P. (1963) *Invitation to Sociology*, London: Pelican.

Bermingham, A. (1986) *Landscape and Ideology: The English Rustic Tradition 1740–1860*, Berkeley: University of California Press.

Bianchini, F. and Parkinson, M. (eds) (1993) *Cultural Policy and Urban Regeneration: The West European Experience*, Manchester: Manchester University Press.

—— and Schwengel, H. (1991) 'Re-imagining the city', in J. Corner, and S. Harvey (eds) *Enterprise and Heritage: Crosscurrents of National Culture*, London: Routledge.

Binnie, J. (1995) 'Trading places: consumption, sexuality and the production of Queer space', in D. Bell and G. Valentine (eds) *Mapping Desire: Geographies of Sexualities*, London: Routledge.

Blacksell, M. and Gilg, A. (1975) 'Landscape evaluation in practice: the case of south-east Devon', *Transactions of the Institute of British Geographers* 66: 135–40.

Blake, G. (1951) *The Heart of Scotland*, London: Batsford.

Blake, S. L. (1990) 'A woman's trek: what difference does gender make? *Women's Studies International Forum* 13, 4: 347–55.

Blunden, J. and Curry, N. (1988) *A Future for Our Countryside*, Oxford: Basil Blackwell in association with the Countryside Commission.

—— and —— (1989) *A People's Charter?* London: HMSO.

Blunt, A. and Rose, G. (eds) (1994) *Writing Women and Space: colonial and post-colonial geographies*, New York: Guilford.

Bondi, L. (1992a) 'Gender symbols and urban landscapes', *Progress in Human Geography* 16, 2: 157–70.

—— (1992b) 'Gender and dichotomy', *Progress in Human Geography* 16, 1: 98–104.

—— (1998) 'Gender, class and urban space: public and private space in contemporary urban landscape', in C. Aitchison and F. Jordan (eds) *Gender, Space and Identity: Leisure, Culture and Commerce*, Eastbourne: Leisure Studies Association.

Boniface, P. and Fowler, P. J. (1993) *Heritage and Tourism in 'the Global Village'*, London: Routledge.

Bonner, P. and Isles, B. (1995) *Travel Her Way*, Dunstable: Pieces Publishers.

Boorstin, D. J. (1964) *The Image: A Guide to Pseudo-events in America*, New York: Harper & Row.

Bourassa, S. C. (1991) *The Aesthetics of Landscape*, London: Belhaven Press.

Bourdieu, P. (1984) *Distinction: A Social Critique of the Judgement of Taste*, London: Routledge.

Bramham, P. (1989) *Leisure and Urban Processes*, London: Routledge.

Breese, S. (1998) 'In search of Englishness; in search of votes', in J. Arnold, K. Davies and S. Ditchfield (eds) *History and Heritage: Consuming the Past in Contemporary Culture*, Shaftesbury: Donhead.

Brett, D. (1993) 'The construction of heritage', in B. O'Connor, and M. Cronin (eds) *Tourism in Ireland: A Critical Analysis*, Cork: Cork University Press.

Briggs, A. (1962) *William Morris: Selected Writings and Designs*, Harmondsworth: Penguin.

Brooks, A. (1997) *Postfeminisms: Feminism, Cultural Theory and Cultural Forms*, London: Routledge.

Brown, D. (1996) 'Genuine fakes', in T. Selwyn (ed.) *The Tourist Image: Myths and Myth Making in Tourism*, Chichester: Wiley.

Brown, R. M. (1935) 'The business of recreation', *Geographical Review* 25: 467–75.

Bunce, M. (1994) *The Countryside Ideal: Anglo-American Images of Landscape*, London: Routledge.

Burke, E. (1998) *A Philosophical Enquiry into the Origin of our Ideas of the Sublime and the Beautiful*, Oxford: Oxford University Press.

Burl, A. (1979) *Prehistoric Avebury*, London: Yale University Press.

Burnett, P. (1973) 'Social change, the status of women and models of city form and development', *Antipode* 5: 57–61.

Butler, J. (1990) *Gender Trouble: Feminism and the Subversion of Identity*, London: Routledge.

Butler, R. W. (1980) 'The concept of a tourism area cycle of evolution: implications for management of resources', *The Canadian Geographer* 24, 1: 5–12.

—— (1985) 'Evolution of tourism in the Scottish Highlands', *Annals of Tourism Research* 12: 371–91.

—— (1998) 'Tartan mythology: the traditional tourist image of Scotland', in G. Ringer (ed.) *Destinations: Cultural Landscapes of Tourism*, London: Routledge.

—— (1999) 'Sustainable tourism: a state-of the-art review', *Tourism Geographies* 1, 1: 7–25.

—— and Waldbrook, L. A. (1991) 'A new planning tool; the tourist opportunity spectrum', *Journal of Tourism Studies* 2, 1: 2–14.

Butler, T. (1996) '"People like us": the gentrification of Hackney in the 1980s', in T. Butler and M. Rustin (eds) *Rising in the East: The Regeneration of East London*, London: Lawrence & Wishart.

Buzard, J. (1993) *The Beaten Track: European Tourism, Literature, and the Ways to Culture, 1800–1918*, New York: Oxford University Press.

Cano, V. and Prentice, R. (1998) 'Opportunities for endearment to place through electronic "visiting": WWW homepages and the tourism promotion of Scotland', *Tourism Management* 19, 1: 67–73.

Caplan, P. (ed) (1997) *Food, Health and Identity*, London: Routledge.

Carew, R. (1602) *The Survey of Cornwall*, quoted in E. Moir (1964) *The Discovery of Britain: The English Tourists 1540–1840*, London: Routledge & Kegan Paul.

Carlson, A. W. (1980) 'Geographical research on international and domestic tourism', *Journal of Cultural Geography* 1, 1: 149–60.

Carr, E. H. (1964) *What is History?* Harmondsworth: Penguin.

Casson, H. (1975) quoted in British Broadcasting Corporation (ed.) *The Spirit of the Age*, London: British Broadcasting Corporation.

Cherry, G. (1985) 'Scenic heritage and national parks lobbies and legislation in England and Wales', *Leisure Studies* 4: 127–39.

Chiang, Y. (1937) *The Silent Traveller: A Chinese Artist in Lakeland*, London: Country Life.

City Fringe Partnership (1996) *Revitalising the City Fringe: Challenge Fund Proposals for 1996/7*, London: City Fringe Partnership.

—— (1997) *Building on Success: City Fringe Partnership Review 1997*, London: City Fringe Partnership.

Cixous, H. (1983/92) 'The Laugh of the Medusa', in E. Abel and E. K. Abel (eds) *The Signs Reader: Women, Gender and Scholarship*, Chicago: University of Chicago Press.

Clark, K. (1991) *Landscape Into Art*, London: John Murray.

Clarke, G. (1998) 'Crossing borders: lesbian physical education students and their struggles for sexual spaces', paper presented at 'The Big Ghetto: Gender, Sexuality and Leisure', Leisure Studies Association 4th International Conference, July, Leeds Metropolitan University, Leeds.

Clarke, J. and Critcher, C. (1985) *The Devil Makes Work: Leisure in Capitalist Britain*, Basingstoke: Macmillan.

Clifford, J. and Marcus, G. E. (1986) *Writing Culture*, Berkeley: University of California Press.

Coalter, F. (1997) 'Leisure sciences and leisure studies: different concept, same crisis?', *Leisure Sciences* 19: 255–68.

—— (1998) 'Leisure studies, leisure policy and social citizenship: the failure of welfare or the limits of welfare', *Leisure Studies* 17: 21–36.

——, Long, J. and Duffield, B. (1988) *Recreational Welfare: The Rationale for Public Leisure Provision*, Aldershot: Avebury.

Cohen, E. (1979) 'A phenomenology of tourist experiences', *Sociology* 13: 179–201.

Collins, G. and Collins, C. (1981) 'Camillo Sitte Reappraised', in R. Kain (ed.) *Planning for Conservation*, London: Mansell.

Cooper, D. (1994) 'Productive, relational and everywhere? Conceptualising power and resistance within Foucauldian feminism', *Sociology* 28: 435–54.

Coppock, J. T. (1982) 'Geographical contributions to the study of leisure', *Leisure Studies* 1, 1: 1–28.

Corner, J. and Harvey, S. (eds) (1991) *Enterprise and Heritage: Crosscurrents of National Culture*, London: Routledge.

Cornish, V. (1934) 'The scenic amenity of Great Britain', *Geography* 19: 192–202.

Cosgrove, D. E. (1984) *Social Formation and Symbolic Landscape*, London: Croom Helm.

—— (1998) *Social Formation and Symbolic Landscape*, 2nd edn, Wisconsin and London: University of Wisconsin Press.

Countryside Commission (1974) *New Agricultural Landscapes*, Cheltenham: Countryside Commission.

178   *References*

—— (1987) *Policies for Enjoying the Countryside*, Cheltenham: Countryside Commission.

—— (1989) *A Countryside for Everyone*, Cheltenham: Countryside Commission.

—— (1991) *Visitors to the Countryside*, Cheltenham: Countryside Commission.

—— (1995) *Sustainable Rural Tourism: Opportunities for Local Action*, Cheltenham: Countryside Commission.

—— and Sports Council (1986) *Access to the Countryside for Recreation and Sport*, Cheltenham: Countryside Commission.

Countryside Commisssion for Scotland (1978) *Scotland's Scenic Heritage*, Battleby: Countryside Commission for Scotland.

Cox, G. (ed.) (1986) *Hermann Lea, the Hardy Guides: A Guide to the West Country, Volume 1*, Harmondsworth: Penguin.

Cox, J. (1994) *London's East End Life and Traditions*, London: Weidenfeld & Nicholson.

Craig, C. (1982) 'Myths against history: tartanry and kailyard in 19th century Scottish literature', in C. McArthur (ed.) *Scotch Reels: Scotland in Cinema and Television*, London: British Film Institute.

Crang, M., Crang, P. and May, J. (1999) *Virtual Geographies: Bodies, Spaces, Relations*, London: Routledge.

Creese, W. (1966) *The Search for the Environment: The Garden City Before and After*, New Haven: Yale University Press.

Crofts, R. S. and Cooke, R. U. (1974) *Landscape Evaluation: A Comparison of Techniques*, London: University College London, Department of Geography, Occasional Paper no. 25.

Crouch, D. (ed.) (1999) *Leisure/Tourism Geographies: Practices and Geographical Knowledge*, London: Routledge.

Cullen, G. (1971) *The Concise Townscape*, London: The Architectural Press.

Cullingworth J. B. and Nadin, V. (1994) *Town and Country Planning in Britain*, eleventh edition, London and New York: Routledge.

Dames, M. (1996) *The Avebury Cycle*, London: Thames & Hudson.

Daniels, S. (1993) *Fields of Vision: Landscape Imagery and National Identity in England and the United States*, Cambridge: Polity Press.

—— and Cosgrove, D. (eds) (1988) *The Iconography of Landscape: Essays on the Symbolic Representation, Design and Use of Past Environments*, Cambridge: Cambridge University Press.

Dann, G. (1981) 'Tourism motivation', *Annals of Tourism Research* 8: 187–219.

Davidson, P. (1996) 'The holiday and work experiences of women with young children', *Leisure Studies* 15: 89–103.

Davis, M. (1990) *City of Quartz: Excavating the Future in Los Angeles*, Verso: London.

Davis, T. (1987) 'Gay gentrification in the Central (Minneapolis) neighborhood', unpublished research paper, Malcaster College, USA.

de Beauvoir, S. (1949) *The Second Sex*, Harmondsworth: Penguin.

de Certeau, M. (1984) *The Practice of Everyday Life*, Berkeley: University of California Press.

Deacon, B. and Payton, P. (1993) 'Re-inventing Cornwall: culture change on the European periphery', *Cornish Studies*, 2: 1.

Deem, R. (1986) *All Work and No Play? The Sociology of Women and Leisure*, Milton Keynes: Open University Press.

—— (1988) 'Feminism and leisure studies: opening up new directions', in E. Wimbush and M. Talbot (eds) *Relative Freedoms: Women and Leisure*, Milton Keynes: Open University Press.

—— (1996) 'Women, the city and holidays', *Leisure Studies* 15: 105–19.

—— (1999) 'How do we get out of the ghetto? Strategies for research on gender and leisure for the twenty-first century', *Leisure Studies* 18: 161–77.

Defoe, D. (1712) *A Tour Through the Whole Island of Great Britain*, reprinted 1970, Harmondsworth: Penguin.

Denton, P. (ed.) (1988) *Betjeman's London*, London: John Murray.

Department of Culture, Media and Sport (1999) *Arts and Sport: A Report to the Social Exclusion Unit*, London: DCMS.

Department of Environment Transport and the Regions (1998) *Transport Statistics Great Britain, 1998 Edition*, London: Government Statistical Office.

Devon County Council (1988) *The Tarka Project – An Integrated Conservation, Recreation and Tourism Strategy for North Devon*, Exeter: Devon County Council.

Discover Islington and Mazorca Ltd. (1997) *PR and Marketing Strategy for the City Fringe Partnership Cultural Quarters Steering Group*, London: Discover Islington and Mazorca.

Dixey, R. and Talbot, M. (1982) *Women, Leisure and Bingo*, Leeds: Trinity and All Saints College.

Donnelly, P. (1986) 'The paradox of parks: politics of recreational land use before and after the mass trespasses', *Leisure Studies* 5: 211–31.

Dower, J. (1945) *National Parks in England and Wales*, London: HMSO.

Dower, M. (1965) *The Fourth Wave*, London: Civic Trust.

Duffield, B. S. and Owen, M. L. (1970) *Leisure + Countryside = A Geographical Appraisal of Countryside Recreation in Lanarkshire*, Edinburgh: Tourism and Recreation Research Unit, University of Edinburgh.

Duncan, N. (ed.) (1996) *BodySpace: Destabilising Geographies of Gender and Sexuality*, London: Routledge.

Durie, A. J. (1994) 'The development of Scotland as a tourist destination', in A. V. Seaton (ed.) *Tourism: State of the Art*, Chichester: Wiley.

Dworkin, A. (1981) *Our Blood: Prophesies and Discourses on Sexual Politics*, London: Women's Press.

Edensor, T. and Kothari (1994) 'The masculinisation of Stirling's heritage', in V. Kinnaird and D. Hall (eds) (1994) *Tourism: A Gender Analysis*, Chichester: Wiley.

Edwards, E. (1996) 'Postcards: greetings from another world', in T. Selwyn (ed.) *The Tourist Image: Myths and Myth Making in Tourism*, Chichester: Wiley.

Elshtain, J. B. (1981) *Public Man, Private Woman*, Princeton, NJ: Princeton University Press.

Emerick, K. (1998) 'Sir Charles Peers and after: from frozen monuments to fluid landscapes', in J. Arnold, K. Davies and S. Ditchfield (eds) *History and Heritage: Consuming the Past in Contemporary Culture*, Shaftesbury: Donhead.

English Heritage (1998) *Conservation-led Regeneration: The Work of English Heritage*, London: English Heritage.

English Tourist Board (1989) *Tourism for All*, report of the working party, London: English Tourist Board.

Enloe, C. (1989) *Bananas, Beaches and Bases: Making Feminist Sense of International Politics*, London: Pandora.

Featherstone, M. (1991) *Consumer Culture and Postmodernism*, London: Sage.

Fees, C. (1996) 'Tourism and the politics of authenticity in a North Cotswold town' in T. Selwyn (ed.) *The Tourist Image: Myths and Myth Making in Tourism*, Chichester: Wiley.

Feifer, M. (1985) *Going Places: The Ways of the Tourist from Imperial Rome to the Present Day*, London: Macmillan.

Foucault, M. (1976) *The Birth of the Clinic*, London: Tavistock.

—— (1977) *Discipline and Punish: The Birth of the Prison*, Harmondsworth: Peregrine.

Fowler, D. D. and Hardesty, D. L. (1994) *Others Knowing Others: Perspectives on Ethnographic Careers*, Washington, DC: Smithsonian Institute Press.

Fowler, P. J. (1992) *The Past in Contemporary Society: Then, Now*, London: Routledge.

Fyfe, N. R. (1998) *Images Of The Street: Planning, Identity and Control in Public Space*, London: Routledge.

—— and Bannister, J. (1998) 'Eyes upon the street', in N. R. Fyfe (ed.) *Images Of The Street: Planning, Identity and Control in Public Space*, London: Routledge.

Garcia-Ramon, M. D., Castener, M. and Centelles, N. (1988) 'Women and geography in Spanish universities', *Professional Geographer* 40: 307–15.

Gibson, H. and Jordan, F. (1998) 'A cross-national study of solo women travellers in the United States and the United Kingdom', paper presented at ' The Big Ghetto: Gender, Sexuality and Leisure', Leisure Studies International Conference, Leeds Metropolitan University, Leeds, July 1998.

Giddens, A. (1984) *The Constitution of Society: Outline of a Theory of Structuration*, Cambridge: Polity Press.

—— (1985) *A Contemporary Critique of Historical Materialism, Vol. 2: The Nation-State and Violence*, Cambridge: Polity Press.

Gilbert, E. W. (1939) 'The growth of inland and seaside health resorts in England', *Scottish Geographical Magazine* 55, 16–35.

Gilpin, W. (1782) *Observations on the River Wye, and Several Parts of South Wales etc, Relative Chiefly to Picturesque Beauty; Made in the Summer of the Year 1770*, London.

—— (1786) *Observations Relative Chiefly to Picturesque Beauty, Made in the Year 1772 on Several Parts of England, Particularly the Mountains, and Lakes of Cumberland and Westmoreland*, Vol. 2, London.

Glaser, J. R. and Zenetou, A. A. (eds) (1994) *Gender Perspectives: Essays on Women in Museums*, Washington, DC: Smithsonian Institution Press.

Gleeson, B. (1999) *Geographies of Disability*, London: Routledge.

Gluckman, A. and Reed, B. (eds) (1997) *Homo Economics: Capitalism, Community and Lesbian and Gay Life*, New York: Routledge.

Glyptis, S. (1991) *Countryside Recreation*, Harlow: Longman/ILAM.

Gold, M. (1984) 'A history of nature', in D. Massey and J. Allen (eds) *Geography Matters! A Reader*, Cambridge: Cambridge University Press.

Gold, J. and Gold, M. (1994) ' "Home at Last!": building societies, home ownership and the imagery of English suburban promotion in the interwar years', in

J. Gold and S. Ward (eds) *Place Promotion: The Use of Publicity and Marketing to Sell Towns and Regions*, Chichester: Wiley.

—— and —— (1995) *Imagining Scotland: Tradition, Representation and Promotion in Scottish Tourism since 1750*, Aldershot: Scolar.

Goldman, M. F. and Papson, S. (1998) *Nike Culture: The Sign of the Swoosh*, London: Sage.

Goodey, B. (1994a) 'Art-full places: public art to sell public spaces?', in J. Gold, and S. Ward (eds) *Place Promotion: The Use of Publicity and Marketing to Sell Towns and Regions*, Chichester: Wiley.

—— (1994b) 'Spreading the benefits of heritage visitor quarters', *International Journal of Heritage Studies* 1, 1: 18–29.

Gosling, P. (1999) Walcott on Tourism, *Barbados Sunday Advocate*, 23 May: 19.

Graham-Dixon, A. (1996) *A History of British Art*, London: BBC Books.

Gramsci, A. (1985) *Prison Notebooks: Selections*, New York: International.

Grant, E. (1988) 'The sphinx in the north: Egyptian influences on landscape, architecture and interior design in eighteenth and nineteenth century Scotland', in D. Cosgrove and S. Daniels (eds) *The Iconography of Landscape: Essays on the Symbolic Representation, Design and Use of Past Environments*, Cambridge: Cambridge University Press.

Granville, A. B. (1841) *The Spas of England, and Principal Sea-bathing Places*, reprinted 1971, Bath: Adams & Dart.

Green, B. (1985) *Countryside Conservation*, London: E. & F. N. Spon.

Green, E. (1998) 'Women doing friendship: an analysis of women's leisure as a site of identity construction, empowerment and resistance', *Leisure Studies* 17, 3: 171–205.

——, Hebron, S. and Woodward, D. (1987) *Leisure and Gender: A Study of Sheffield Women's Leisure Experiences*, London: The Sports Council/Economic and Social Research Council.

——, ——, and —— (1990) *Women's Leisure, What Leisure?*, London: Macmillan.

Gregory, D. (1994) *Geographical Imaginations*, Oxford: Basil Blackwell.

—— and Urry, J. (eds) (1985) *Social Relations and Spatial Structures*, Macmillan: Basingstoke.

——, Martin, R. and Smith, G. (1994) *Human Geography: Society, Space and Social Science*, Basingstoke: Macmillan.

Grimsley, R. (1973) *The Philosophy of Rousseau*, Oxford: Oxford University Press.

Gruffudd, P. (1994) 'Selling the countryside: representations of rural Britain', in J. Gold and S. Ward (eds) *Place Promotion: The Use of Publicity and Marketing to Sell Towns and Regions*, Chichester: Wiley.

Guardian (1999) 'Avebury stones daubed in GM protest', *Guardian*, 19 June.

—— (1999) 'Solstice sun rises on anarchy', *Guardian*, 22 June.

Habermas, J. (1989) *The New Conservatism: Cultural Criticism and the Historians' Debate*, Cambridge: Polity Press.

Halfacree, K. (1997) 'Contrasting roles for the post-productivist countryside: a postmodern perspective on counterurbanisation', in P. Cloke and J. Little (eds) *Contested Countryside Cultures: Otherness, Marginalisation and Rurality*, London: Routledge.

Hall, P. (1988) 'Industrial revolution in reverse?', *The Planner*, January: 15–19.

Hall, T. (1998) *Urban Geography*, London: Routledge.

Hannigan, J. (1998) *Fantasy City: Pleasure and Profit in the Postmodern Metropolis*, London: Routledge.

Haraway, D. (1985) 'Class, race, sex, scientific objects of knowledge: a socialist feminist perspective on the construction of productive knowledge and some political consequences', in V. Haas and C. Perucci (eds) *Women in Scientific and Engineering Professions*, Michigan: University of Michigan Press.

Hardy, F. (1990) *Scotland in Film*, Edinburgh: Edinburgh University Press.

Harley, J. B. (1988) 'Maps, knowledge and power', in D. Cosgrove and S. Daniels (eds), *The Iconography of Landscape*, Cambridge: Cambridge University Press.

Harrison, C. (1991) *Countryside Recreation in a Changing Society*, London: The T.M.S. Partnership Ltd.

——, Limb, M. and Burgess, J. (1986) 'Recreation 2000: views of the country from the city', *Landscape Research* 11, 19–24.

Hartsock, N. (1990) 'Foucault on power', in L. Nicholson (ed.), *Feminism/Postmodernism*, London: Routledge.

Harvey, D. (1969) *Explanation in Geography*, London: Edward Arnold.

—— (1973) *Social Justice and the City*, London: Edward Arnold.

—— (1985a) *The Urbanisation of Capital*, Oxford: Basil Blackwell.

—— (1985b) *Consciousness and the Urban Experience*, Oxford: Basil Blackwell.

—— (1989) *The Condition of Postmodernity*, Oxford: Blackwell.

—— (1989a) *The Urban Experience*, Oxford: Basil Blackwell.

—— (1989b) 'Down towns', *Marxism Today*, January.

Haywood, L. (ed.) (1994) *Community Leisure and Recreation*, Oxford: Butterworth-Heinemann.

——, Kew, F. and Bramham, P. (1989) *Understanding Leisure*, London: Hutchinson.

Hazlitt, W. (1814) 'On the love of the country' in *The Examiner*.

—— (1822) 'On Going a Journey', reprinted (1948) in G. Keynes (ed.), *Selected Essays of William Hazlitt 1778: 1830*, London: Nonesuch.

Heal, F. (1990) *Hospitality in Early Modern England*, Oxford: Clarendon.

Health Promotion Research Trust (1987) *Women's Leisure and Well-being: Interim Paper*, Edinburgh: Centre for Leisure Research, Moray House College of Education.

Hebbert, M. (1998) *London: More by Fortune than Design*, Chichester: Wiley.

Heidegger, M. (1980) *An Introduction to Metaphysics*, New Haven: Yale University Press.

Helsinger, E. (1994) 'Turner and the representation of England' in W. Mitchell (ed.) *Landscape and Power*, Chicago and London: University of Chicago Press.

Henderson, K. (1994) 'Perspectives on analysing gender, women and leisure', *Journal of Leisure Research* 26, 2: 119–37.

—— and Bialeschki, D. (1991) 'A sense of entitlement to leisure as constraint and empowerment for women', *Leisure Sciences* 13, 1: 51–65.

——, —— and Shaw, S. (1996) *Both Gains and Gaps: Feminist Perspectives on Women's Leisure*, Pennsylvania: Venture Publishing.

——, ——, —— and Freysinger, V. (1989) *A Leisure of One's Own: A Feminist Perspective in Women's Leisure*, Pennsylvania, USA: Venture Publishing.

Henry, I. (1997) 'The politics of sport and symbolism in the city: a case study of the Lyon conurbation', *Managing Leisure* 2: 65–81.

Herbert, D. T. (1995) 'Heritage as literary place', in D. T. Herbert (ed.) *Heritage, Tourism and Society*, London and New York: Mansel.

Hetherington, K. (1996) 'The utopics of social ordering – Stonehenge as a museum without walls', in S. Macdonald and G. Fyfe (eds) *Theorizing Museums*, Oxford: Blackwell.

Hetherington, P. (1999) 'The Easel of the North aims to outsoar the Angel of the North', in *Guardian*, 17 June.

Hewison, R. (1987) *The Heritage Industry: Britain in a Climate of Decline*, London: Methuen.

Hills, P. (1994) 'The cultural potency of Scotland', in J. M. Fladmark (ed.) *Cultural Tourism*, London: Donhead.

Hitchcock, M., Stanley, N. and King Chung, S. (1997) 'The south-east Asian "living museum" and its antecedents', in S. Abram, J. Waldren and D. V. L. Macleod (eds) *Tourists and Tourism: Identifying with People and Places*, Oxford: Berg.

Hobhouse, Lord Justice (1947) *Paper of the National Parks Committee (England and Wales)*, London: HMSO.

Holcomb, B. and Luongo, M. (1996) 'Gay tourism in the United States', *Annals of Tourism Research*, 711–13.

Honour, H. (1979) *Romanticism*, London: Allen Lane.

Hooper-Greenhill, E. (ed.) (1997) *Cultural Diversity: Developing Museum Audiences in Britain*, London: Leicester University Press/Cassell.

Horst, O. H. (1981) *Papers in Latin American Geography in Honor of Lucia C. Harrison*, Muncie, IA: Conference of Latin American Geographers, special publication no. 1.

Howard, E. (1898) *Tomorrow: A Peaceful Path to Real Reform*, London: Swan & Sonnenshein.

Hudson, K. (1987) *Museums of Influence*, Cambridge: Cambridge University Press.

Hughes, G. (1992) 'Tourism and the geographical imagination', *Leisure Studies* 11, 1: 31–42.

—— (1998) 'Tourism and the semiological realisation of space', in G. Ringer (ed) *Destinations: Cultural Landscapes of Tourism*, London: Routledge.

Hughes, H. (1997) 'The significance and effects of the use on non-local urban space by gay men', paper presented at 'Leisure, Culture and Commerce', Annual Conference of the UK Leisure Studies Association, Roehampton Institute London, September.

Humberstone, B. (1997) 'Challenging dominant ideologies in the research process', in G. Clarke and B. Humberstone (eds) *Researching Women and Sport*, Basingstoke: Macmillan.

Humphreys, R. (1997) *London, the Rough Guide*, London: Rough Guides.

Ihse, M. and Norderhaug, A. (1995) 'Biological values of the Nordic cultural landscape: different perspectives', *International Journal of Heritage Studies* 1, 3: 156–70.

Inwood, S. (1998) *A History of London*, London: Macmillan.

Itzin, C. (ed.) (1992) *Pornography, Women, Violence and Civil Liberties*, Oxford: Oxford University Press.

Jackson, P. and Smith, S. J. (1984) *Exploring Social Geography*, London: Allen & Unwin.

Jacques, D. (1995) 'The rise of cultural landscapes', *International Journal of Heritage Studies* 1, 2: 91–101.

Jeffreys, S. (1999) 'Globalising sexual exploitation: sex tourism and the traffic in women', *Leisure Studies* 18, 3: 179–96.

Jenkins, J. and James, P. (1994) *From Acorn to Oak Tree: The Growth of the National Trust 1895–1994*, Basingstoke: Macmillan.

Joad, C. E. M. (1938) 'The people's claim', in C. Williams-Ellis (ed.) *England and the Beast*, London: Readers' Union and J. Dent & Sons.

Johnston, R. J. (1984) *City and Society*, second edition, London: Hutchinson.

Jordan, F. and Gibson, H. (1998) 'Travelling solo: a cross-cultural study of British and American women aged 30–50', paper presented at 'The Big Ghetto: Gender, Sexuality and Leisure', International Conference of the UK Leisure Studies Association, Leeds Metropolitan University, July (copy available from the authors).

Kappeler, S. (1986) *The Pornography of Representation*, Cambridge: Polity.

Kay, T. (1996) 'Women's work and women's worth: the leisure implications of women's changing employment patterns', *Leisure Studies* 15, 1: 49–64.

Kinnaird, V. and Hall, D. (1994) *Tourism: A Gender Analysis*, Chichester: Wiley.

Knopp, L. (1995) 'Sexuality and urban space', in D. Bell and G. Valentine (eds) *Mapping Desire: Geographies of Sexualities*, London: Routledge.

—— (1997) 'Gentrification and gay neighbourhood formation in New Orleans', in A. Gluckman and B. Red (eds) *Homo Economics: Capitalism, Community and Lesbian and Gay Life*, New York: Routledge.

Kroeber, K. (1975) *Romantic Landscape Vision, Constable and Wordsworth*, Madison: University of Wisconsin Press.

Larkham, P. (1996) *Conservation and the City*, London and New York: Routledge.

Lash, S. (1990) *Sociology of Postmodernism*, London: Routledge.

Lauria, M. and Knopp, L. (1985) 'Towards an analysis of the role of gay communities in urban renaissance', *Urban Geography* 6: 152–69.

Law, C. M. (1994) *Urban Tourism: Attracting Visitors to Large Cities*, London: Mansell.

Le Corbusier (1933) *The Radiant City*, reprinted in 1967, London: Faber & Faber.

Leask, A. and Yeoman, I. (eds) (1999) *Heritage Visitor Attractions: An Operations Management Perspective*, London: Cassell.

Leneman, L. (1987) 'The effects of Ossian on Lowland Scotland', in J. J. Carter and J. H. Pittock (eds) *Aberdeen and Enlightenment*, Aberdeen: Aberdeen University Press.

Ley, D. (1983) *A Social Geography of the City*, New York: Harper & Row.

Local Government Association (1999) *Making The Case For Leisure: Research Specification*, London Local Government Association.

London Borough of Tower Hamlets (1996) *Eastside: Challenge Fund Submission*, London.

Losch, A. (1940) *Die Raumliche Ordnung der Wirtschaft (The Economics of Location)* translated by W. H. Woglom and W. F. Stolper, 1954, New Haven: Yale University Press.

Lowe, P. and Goyder J. (1983) *Environmental Groups in Politics*, London: Allen & Unwin.

Lowenthal, D. (1982) 'Revisiting valued landscapes', in J. Gold and J. Burgess (eds) *Valued Environments*, London: Allen & Unwin.

—— (1985) *The Past is a Foreign Country*, Cambridge: Cambridge University Press.

—— and Prince, H. (1965) 'English landscape tastes', *Geographical Review* 55: 187–222.

Lévi-Strauss, C. (1968) *Totemism*, London: Penguin.

Ley, D. (1983) *A Social Geography of the City*, New York: Harper & Row.

Lumley, R. (1994) 'The debate on heritage reviewed', in R. Miles and L. Zavala (eds) *Towards the Museum of the Future: New European Perspectives*, London: Routledge.

MacCannell, D. (1976) *The Tourist: A New Theory of the Leisure Class*, London: Macmillan.

—— (1992) *Empty Meeting Grounds: The Tourist Papers*, London: Routledge.

McClintock, A. (1995) *Imperial Leather: Race, Gender and Sexuality in the Imperial Context*, London: Routledge.

McCrone, D. (1996) *Understanding Scotland: the Sociology of a Stateless Nation*, London: Routledge.

——, Morris, A. and Kiely, R. (1995) *Scotland the Brand: The Making of Scottish Heritage*, Edinburgh: Edinburgh University Press.

McDonald, D. and Tungatt, M. (1992) *Community Development and Sport*, London: Sports Council.

McDowell, L. (1993a) 'Space, place and gender relations, part I: feminist empiricism and the geography of social relations', *Progress in Human Geography* 17, 2: 157–79.

—— (1993b) 'Space, place and gender relations, part II: identity, difference, feminist geometries and geographies', *Progress in Human Geography* 17, 3: 305–18.

—— and Sharp, J. (eds) (1997) *Space, Gender, Knowledge: Feminist Readings*, London: Arnold.

McKean, C. (1985) *Stirling and The Trossachs*, Edinburgh: Royal Incorporation of architects in Scotland/Scottish Academic Press.

MacKellar Goulty, S. (1993) *Heritage Gardens: Care, Conservation, Management*, London: Routledge.

MacKinnon, C. (1995) 'Sexuality, pornography and method', in N. Tuana and R. Tong (eds) *Feminism and Philosophy*, Oxford: Westview Press.

Malone, C. (1989) *Avebury*, London: English Heritage/Batsford.

Malraux, A. (1978) 'Museum without walls', in *The Voices of Silence*, trans. S. Gilbert, Princeton, NJ: Princeton University Press.

Massey, D. (1984) *Spatial Divisions of Labour*, London: Macmillan.

—— (1994) *Space, Place and Gender*, Cambridge: Polity.

—— and Jess, P. (eds) (1995) *A Place in the World? Places, Cultures and Globalisation*, Milton Keynes: The Open University.

Mead, G. M. (1934) *Mind, Self and Society*, Chicago: Chicago University Press.

Meades, J. (1997) 'Roll away the stones', *Observer*, 17 August.

Memmi, A. (1967) *The Coloniser and the Colonised*, Boston: Beacon Press.

Merriman, N. (1991) *Beyond the Glass Case: The Past, the Heritage and the Public in Britain*, Leicester: Leicester University Press.

Miles, M. (1997) *Art Space and the City: Public Art and Urban Futures*, London: Routledge.

Miller, D., Jackson, P., Thrift, N., Holbrook, B. and Rowlands, M. (1998) *Shopping, Place and Identity*, London: Routledge.

Miller, M. (1981) 'Raymond Unwin 1863–1940' in G. Cherry (ed.) *Pioneers in British Town Planning*, London: Architectural Press.

Mills, S. (1991) *Discourses of Difference: an analysis of women's travel writing and colonialism*, London: Routledge.

—— (1998) 'Post-colonial feminist theory', in S. Jackson and J. Jones (eds) *Contemporary Feminist Theories*, Edinburgh: Edinburgh University Press.

Minister of Land and Natural Resources and Secretary of State for Wales (1966) *Leisure in the Countryside: England and Wales*, London: HMSO.

Mitchell, D. (1992) 'Iconography and locational conflict from the underside: free speech, people's park and the politics of homelessness', *Political Geography* 11: 152–69.

Mitchell, J. W. T. (1994) 'Imperial landscape', in J. W. T. Mitchell (ed.) *Landscape and Power*, Chicago and London: University of Chicago Press.

Mitchell, L. S. and Murphy, P. E. (1991) 'Geography and tourism', *Annals of Tourism Research* 18, 1: 57–70.

Moir, E. (1964) *The Discovery of Britain: The English Tourists 1540–1840*, London: Routledge & Kegan Paul.

Moissec, J. M. (1977) 'Un modèle de l'espace touristique', *L'Espace Geographique* 6: 41–8.

Momaas, H. (1997) 'European leisure studies at the crossroads? A history of leisure research in Europe', *Leisure Sciences* 19, 4: 241–54.

Morris, C. (ed.) (1982) *The Illustrated Journeys of Celia Fiennes 1685–circa 1712*, London and Sydney: Webb & Bower.

Mort, F. (1996) *Cultures of Consumption: Masculinities and Social Space in Late Twentieth Century Britain*, London: Routledge.

Mowl, G. and Turner, J. (1995) 'Women, gender, leisure and place: towards a more "humanistic" geography of women's leisure', *Leisure Studies* 14, 2: 102–16.

Mumford, L. (1968) *The Urban Prospect*, London: Secker and Warburg.

Munt, S. (1995) 'The lesbian *flaneur*', in D. Bell and G. Valentine (eds) *Mapping Desire: Geographies of Sexualities*, London: Routledge.

Myslik, W. (1996) 'Renegotiating the social/sexual identities of places: gay communities as safe havens or sites of resistance?', in N. Duncan (ed.) *BodySpace: Destabilising Geographies of Gender and Sexuality*, London: Routledge.

Nast, H. and Pile, S. (eds) (1998) *Places Through the Body*, London: Routledge.

Nicholson, M. (1959) *Mountain Gloom and Mountain Glory: The Development of the Aesthetics of the Infinite*, Seattle and London: University of Washington Press.

Nietzsche, F. (1980) *On the Advantage and Disadvantage of History for Life*, Indianapolis: Hackett.

Osborne, B. S. (1988) 'The iconography of nationhood in Canadian art', in D. Cosgrove and S. Daniels (eds) *The Iconography of Landscape: Essays on the Symbolic Representation, Design and Use of Past Environments*, Cambridge: Cambridge University Press.

Ousby, I. (1990) *The Englishman's England: Taste, Travel and the Rise of Tourism*, Cambridge: Cambridge University Press.

Pahl, R. E. (1970) *Whose City?* London: Longman.

Parker, S. (1971) *The Future of Work and Leisure*, London: MacGibbon & Kee.

—— (1983) *Leisure and Work*, London: Allen & Unwin.

—— (1988) 'A conventional leisure theory', in A. Clarke and B. Bacon (eds) *Leisure Theory: Four Perspectives*, Sheffield: University of Sheffield.

Parks, G. B. (1964) 'The turn to the romantic in the travel literature of the eighteenth century', *Modern Languages Quarterly* 25: 22–33.

Patmore, J. A. (1972) *Land and Leisure*, London: Pelican.

—— (1983) *Recreation and Resources*, London: Blackwell.

Peake, L. (ed.) (1989) 'The challenge of feminist geography', *Journal of Geography in Higher Education* 13: 85–121.

Pearce, S. (1994) 'Studying museum material and collections', *International Journal of Heritage Studies* 1, 1: 30–4.

Penrose, J. (1993) 'Reification in the name of change: the impact of nationalism on social constructions of nation, people and place in Scotland and the United Kingdom', in P. Jackson and J. Penrose (eds) *Constructions of Race, Place and Nation*, London: UCL Press.

Pickard, R. (1996) *Conservation in the Built Environment*, Harlow: Longman.

Pimlott, J. A. R. (1947) *The Englishman's Holiday: A Social History*, London: Faber & Faber.

Plog, S. (1974) 'Why destinations rise and fall in popularity', *The Cornell Hotel and Restaurant Administration Quarterly* 14, 4: 55–8.

Pocock, D. C. D. (1981) 'Imaginative literature and the geographer', in D. C. D. Pocock (ed.) *Humanistic Geography and Literature: Essays on the Experience of Place*, London and Sydney: Croom Helm.

Porter, D. (ed.) (1997) *Internet Culture*, London: Routledge.

Porter, R. (1996) *London: A Social History*, Harmondsworth: Penguin.

Prentice, R. (1994) 'Heritage: a key sector of the "new" tourism', in C. P. Cooper and A. Lockwood (eds) *Progress in Tourism, Recreation and Hospitality Management Volume 5*, Chichester: Wiley.

Priestley, J. B. (1935) *The Beauty of Britain*, London: Batsford.

Prince, H. (1988) 'Art and agrarian change, 1710–1815', in D. Cosgrove, and S. Daniels (eds) *The Iconography of Landscape: Essays on the Symbolic Representation, Design and Use of Past Environments*, Cambridge: Cambridge University Press.

Pringle, T. R. (1988) 'The privation of history: Landseer, Victoria and the Highland Myth', in D. Cosgrove and S. Daniels (eds) *The Iconography of Landscape: Essays on the Symbolic Representation, Design and Use of Past Environments*, Cambridge: Cambridge University Press.

Pritchard, A., Morgan, N., Sedgley, D. and Jenkins, A. (1998) 'Gay tourism destinations: identity, sponsorship and degaying', in C. Aitchison and F. Jordan (eds) *Gender, Space and Identity: Leisure, Culture and Commerce*, Eastbourne: Leisure Studies Association.

Pugin, A. (1836) *Contrasts: Or a Parallel between the Noble Edifices of the Middle Ages and Corresponding Buildings of the Present Day, Shewing the Present Decay of Taste'*, London.

Punter, D. (1994) 'The picturesque and the sublime: two worldscapes', in S. Copley and P. Garside (eds) *The Politics of the Picturesque*, Cambridge: Cambridge University Press.

Rapoport, R. and Rapoport, R. N. (1975) *Leisure and the Family Life Cycle*, London: Routledge & Kegan Paul.

Ratcliffe, R. U. (1943) *Urban Land Economics*. Westport, CN: Greenwood Press.

Reed, M. (1997) *The Landscape of Britain: From the Beginnings to 1914*, London: Routledge.

Relph, E. (1976) *Place and Placelessness*, London: Pion.

Rendell, J. (1998) 'Displaying sexuality: gendered identities and the early nine-teenth-century street', in N. R. Fyfe (ed.) *Images Of The Street: Planning, Identity and Control in Public Space*, London: Routledge.

Richards, G. (1995) 'Politics of national tourism policy in Britain', *Leisure Studies* 14: 153–73.

—— (1996) *Cultural Tourism in Europe*, Wallingford: CAB International.

Richards, J. and MacKenzie, J. (1986) *The Railway Station: A Social History*, Oxford and New York: Oxford University Press.

Richardson, J. (1995) *London and its People: A Social History From Medieval Times to the Present Day*, London: Barrie & Jenkins.

Riley, R. W. (1994) 'Movie-induced tourism', in A. V. Seaton (ed.) *Tourism: The State of the Art*, Chichester: Wiley.

Robb, J. G. (1998) 'Tourism and legends: archaeology of heritage', *Annals of Tourism Research* 25, 3: 579–96.

Roberts, K. (1978) *Contemporary Society and the Growth of Leisure*, London: Longman.

Robins, K. (1991) 'Tradition and translation: national culture in its global context', in J. Corner and S. Harvey (eds) *Enterprise and Heritage: Crosscurrents of National Culture*, London and New York: Routledge.

Rogers, M. F. (1998) *Barbie Culture*, London: Sage.

Rogers, R. (1992) *Architecture: A Modern View*, New York: Thames & Hudson.

Rojek, C. (1988) *Leisure for Leisure*, London: Sage.

—— (1993) 'After popular culture: hyper-reality and leisure', *Leisure Studies* 12: 277–89.

—— (1995) *Decentring Leisure: Rethinking Leisure Theory*, London: Sage.

—— and Urry, J. (1997) *Touring Cultures: Transformations of Travel and Theory*, London: Routledge.

—— and —— (1998) 'Cybertourism and the phantasmagoria of place', in G. Ringer (ed.) *Destinations: Cultural Landscapes of Tourism*, London: Routledge.

Rose, G. (1993) *Feminism and Geography: The Limits of Geographical Knowledge*, Cambridge: Polity Press.

—— (1995) 'Place and identity: a sense of place', in D. Massey and P. Jess (eds) *A Place in the World? Places, Cultures and Globalization*, Milton Keynes: Open University Press.

—— (1996) 'As if the mirrors had bled: masculine dwelling, masculinist theory and feminist masquerade', in N. Duncan (ed.) *BodySpace: Destabilising Geographies of Gender and Sexuality*, London: Routledge.

Ross, M. (1991) *Planning and the Heritage: Policy and Procedures*, London: Spon.

Rossi, A. (1982) *The Architecture of the City*, Massachusetts: The M. I. T. Press.

Rothenberg, T. (1995) '"And she told two friends": lesbians creating urban space', in D. Bell and G. Valentine (eds) *Mapping Desire: Geographies of Sexualities*, London: Routledge.

Roy, A. (1997) *The God of Small Things*, London: Flamingo.

Rubinstein, D. (1977) 'Cycling in the 1890s', *Victorian Studies* 1: 1.

Rule, J. (1973) *Private Lives, Public Surveillance*, London: Allen-Lane.

Ruskin, J. (1849) *The Seven Lamps of Architecture*, reprinted 1961, New York: Noonday Press.

Said, E. (1978) *Orientalism*, London: Routledge.

—— (1993) *Culture and Imperialsim*, London: Chatto & Windus.

Samdhal, D. and Jekubovich, N. J. (1997) 'A critique of leisure constraints: comparative analysis and understanding', *Journal of Leisure Research* 29, 4: 430–52.

Schafernich, S. M. (1996) 'Two heritage centres in England', *Museum Management and Curatorship* 15, 1: 37–47.

Schama, S. (1995) *Landscape and Memory*, London: HarperCollins.

Schivelbusch, W. (1986) *The Railway Journey, Trains and Travel in the Nineteenth Century*, Oxford: Blackwell.

Schofield, P. (1996) 'Cinematographic images of a city: alternative heritage tourism in Manchester', *Tourism Management* 17, 5: 333–40.

Schouten, F. F. J. (1995) 'Heritage as historical reality', in D. T. Herbert (ed.) *Heritage, Tourism and Society*, London: Mansell.

Scott, Lord Justice (1942) *Paper of the Committee on Land Utilisation in Rural Areas*, Ministry of Works and Planning, London: HMSO.

Scott, P. H. (1994) 'The image of Scotland in literature', in J. M. Fladmark (ed.) *Cultural Tourism*, London: Donhead.

Scottish Tourist Board (*c*.1995) 'Film tourism: business guidelines for the tourism industry', Edinburgh: Scottish Tourist Board.

Scraton, S. (1994) 'The changing world of women and leisure: "post-feminism" and "leisure" ', *Leisure Studies* 13: 249–61.

Seaton, A. V. (1998) 'The history of tourism in Scotland: approaches, sources and issues', in R. MacLellan and R. Smith (eds) *Tourism in Scotland*, London: International Thompson Business Press.

Selby, M. and Morgan, N. (1996) 'Reconstructing place image: a case study of its role in destination market research', *Tourism Management* 17, 4.

Selwyn, T. (1996) *The Tourist Image: Myths and Myth Making in Tourism*, Chichester: Wiley.

Sharp, J. (1996) 'Gendering nationhood: a feminist engagement with national identity' in N. Duncan (ed.) *BodySpace: destabilising geographies of gender and sexuality*, London: Routledge.

Shaw, G. and Williams, A. (1991) 'From bathing hut to theme park: tourism development in South West England', *Journal of Regional Studies* 11, 1: 42.

Shaw, S. (1993) *Transport: Strategy and Policy*, Oxford: Blackwell.

Shields, R. (1991) *Places on the Margin: Alternative Geographies of Modernity*, London: Routledge.

Shoard, M. (1982) 'The Lure of the Moors', in J. Burgess and J. Gold (eds) *Valued Environments*, London: Allen & Unwin.

—— (1987) *This Land is Our Land*, London: Paladin.

Shurmer-Smith, P. and Hannam, K. (1994) *Worlds of Desire, Realms of Power: A Cultural Geography*, London: Arnold.

Sibley, D. (1995) *Geographies of Exclusion*, London: Routledge.

Sinclair, T. (1997) (ed.) *Gender, Work and Tourism*, London: Routledge.

Sitte, C. (1965) *City Planning According to Artistic Principles*, translated by G. Collins and C. Collins, London: Phaidon Press and New York: Random House.

Skeggs, B. (1999) 'Matter out of place: visibility and sexualities in leisure spaces', *Leisure Studies* 18, 3: 213–32.

Skelton, T. and Valentine, G. (eds) (1997) *Cool Places: Geographies of Youth Cultures*, London: Routledge.

Smyth, H. (1994) *Marketing the City: The Role of Flagship Developments in Urban Regeneration*, London: Spon.

Social Exclusion Unit (1998) *Bringing Britain Together: A National Strategy for Neighbourhood Renewal*, London: Social Exclusion Unit.

Soja, E. (1985) 'The spatiality of social life: towards a transformative retheorisation', in D. Gregory and J. Urry (eds) *Social Relations and Spatial Structures*, Basingstoke: Macmillan.

Sontag, S. (1973) *On Photography*, Harmondsworth: Penguin.

Spalding, F. (1986) *British Art Since 1900,* London: Thames & Hudson.

Spivak, G. C. (1985) 'Three women's texts and a critique of imperialism', *Critical Inquiry* 12, 1: 243–62.

—— (1987) *In Other Worlds: Essays in Cultural Politics*, London: Routledge.

—— (1993) *Outside in the Teaching Machine*, London: Routledge.

Spoonley, P. (1995) 'The challenges of post-colonialism', *Sites* 30: 48–68.

Stamp, L. D. (1948) *The Land of Britain: Its Use and Misuse*, London: Longmans.

Stanley, L. (1980) *The Problem of Women and Leisure: An Ideological Construct and a Radical Feminist Alternative*, London: Sports Council/Social Science Research Council.

Stephenson, T. (1989) 'Forbidden land: the struggle for access to mountain and moorland' in Holt, A. *The Ramblers' Association*, Manchester: Manchester University Press.

Stone, P. and Mackenzie, R. (1989) 'Is there an "excluded past" in education?', in D. L. Uzzell (ed.) *Heritage Interpretation: Volume 1 The Natural and Built Environment*, London: Belhaven.

Swain, M. B. (1995) 'Gender in Tourism', *Annals of Tourism Research* 22, 2: 247–66.

Talbot, M. (1979) *Women and Leisure: A State of the Art Review*, London: Sports Council/Social Science Research Council.

Tansley, A. (1939) *The British Isles and their Vegetation*, Cambridge: Cambridge University Press.

Taylor, H. (1997) *A Claim on the Countryside: A History of the British Outdoor Movement*, Edinburgh: Keele University Press.

Taylor, J. (1990) 'The alphabetic universe: photography and the picturesque landscape', in S. Pugh (ed.) *Reading Landscape: Country-City-Capital*, Manchester: Manchester University Press.

Thomas, K. (1983) *Man and the Natural World: Changing Attitudes in England 1500–1800*, Harmondsworth: Penguin.

Thompson, E. P. (1968) *The Making of the English Working Class*, London: Penguin.

Thornton, P. (1993) 'Cornwall and changes in the "tourist gaze"', *Cornish Studies*, second series, no. 1.

Thrift, N. (1989) 'What is culture?', in C. Hamnet, L. McDowell and P. Sarre (eds) *The Changing Social Structure*, London: Sage.

Tiesdell, S., Oc, T. and Heath, T. (1996) *Revitalising Historic Urban Quarters*, Oxford: Architectural Press.

Tomlinson, A. (1981) *Leisure and Social Control*, Brighton: The Chelsea School.

Tooke, N. and Baker, M. (1996) 'Seeing is believing: the effect of film on visitor numbers to screened locations', *Tourism Management* 17, 2: 87–94.

Tourism Geographies (1999) Notes for Authors, *Tourism Geographies*.

Towner, J. (1996) *An Historical Geography of Recreation and Tourism in the Western World 1540–1940*, Chichester: Wiley.

Treuherz, J. (1993) *Victorian Painting*, London: Thames & Hudson.

Trevor-Roper, H. (1983) 'The invention of tradition: the Highland tradition of Scotland', in E. Hobsbawm and T. Ranger (eds) *The Invention of Tradition*, Cambridge: Cambridge University Press.

Tunbridge, J. E. and Ashworth, G. J. (1996) *Dissonant Heritage: the Management of the Past as a Resource in Conflict*, Chichester: Wiley.

Turner, L. and Ash, J. (1975) *The Golden Hordes*, London: Constable.

Unijet (1999) *Winter Brochure 1999/2000*, London: Unijet.

Urban Task Force (1999) *Towards an Urban Renaissance*, London: Department of the Environment Transport and the Regions.

Urry, J. (1990) *The Tourist Gaze*, London: Sage.

—— (1995) *Consuming Places*, Routledge: London.

Valentine, G. (1989) 'A geography of fear', *Area* 21, 4: 385–90.

—— (1993a) '(Hetero)sexing space: lesbian perceptions and experiences in everyday spaces', *Environment and Planning D: Society and Space* 11: 395–413.

—— (1993b) 'Desperately seeking Susan: a geography of lesbian friendships', *Area* 25, 2: 109–16.

—— (1993c) 'Negotiating and managing multiple sexual identities: lesbian time-space strategies', *Transactions of the Institute of British Geographers* 18, 2: 237–48.

—— (1996) '(Re)Negotiating the "heterosexual street": lesbian productions of space' in N. Duncan (ed.) *BodySpace: Destabilising Geographies of Gender and Sexuality*, London: Routledge.

—— (1999) 'A corporeal geography of consumption', *Environment and Planning D: Society and Space* 17, 3: 329–51.

Van den Berge, P. (1994) *The Quest for the Other*, Seattle and London: University of Washington.

Voase, R. (1997) 'The role of flagship cultural projects in urban regeneration: a case study and commentary', *Managing Leisure* 2: 230–41.

Wakeford, R. (1999) 'We want the best for rural England', in *Countryside Focus* 1, April/ May: 1, Cheltenham: Countryside Agency.

Walby, S. (1990) *Theorising Patriarchy*, Oxford: Blackwell.

Wallace, A. (1993) *Walking, Literature, and English Culture: Origins and Uses of Peripatetic in the Nineteenth Century*, Oxford: Clarendon Press.

Walsh, K. (1992) *The Representation of the Past: Museums and Heritage in the Post-modern World*, London: Routledge.

Walvin, J. (1978) *Leisure and Society 1830–1950*, London and New York: Longman.

Ward, S. V. (1998) *Selling Places: The Marketing and Promotion of Towns and Cities 1850–2000*, London: E. & F. N. Spon.

Watson, B. and Scraton, S. (1998) 'Gendered Cities: women and public leisure space in the postmodern city', *Leisure Studies* 17, 2: 123–37.

Watson, J. R. (1970) *Picturesque Landscape and English Romantic Poetry*, London: Hutchinson Educational Ltd.

Wearing, B. (1996) *Gender: The Pleasure and Pain of Difference*, Melbourne, Australia: Longman.

—— (1998) *Leisure and Feminist Theory*, London: Sage.

—— and Wearing, S. (1996) 'Refocussing the tourist experience: the "flaneur" and the "choraster"', *Leisure Studies* 15: 229–44.

Webb, B. and Webb, S. (1913) *English Local Government: the Story of the King's Highway*, London: Longman, Green & Co.

Weber, A. (1909) *Uber den Standort der Industrien (Alfred Weber's Theory of the Location of Industries)*, translated by C. J. Fredrich, 1929, Chicago: University of Chicago Press.

West, T. (1780) *Guide to the Lakes*, second edition, London.

Wilkinson, S. and Kitzinger, C. (1996) *Representing The Other: Feminism and Psychology*, London: Sage.

Williams, R. (1961) *Culture and Society, 1750–1950*, London: Penguin.

—— (1965) *The Long Revolution*, London: Penguin.

—— (1973) *The Country and the City*, London: Hogarth.

—— (1977) *Marxism and Literature*, Oxford: Oxford University Press.

Williams, S. (1998) *Tourism Geography*, London: Routledge.

Williams, W. (1942) *A Book of English Essays*, Harmondsworth: Penguin.

Wilson, E. (1991) *The Sphinx in the City*, London: Virago.

Wimbush, E. (1986) *Women, Leisure and Well-being*, final report to Health Promotion Research Trust, Edinburgh: Centre for Leisure Research, Moray House College of Education.

—— and Talbot, M. (eds) (1988) *Relative Freedoms: Women and Leisure*, Milton Keynes: Open University Press.

Withers, C. (1992) 'The historical creation of the Scottish Highlands', in I. Donnachie and C. Whatley (eds) *The Manufacture of Scottish History*, Edinburgh: Polygon.

Womack, P. (1989) *Improvement and Romance: Constructing the Myth of the Highlands*, Basingstoke and London: Macmillan.

Women and Geography Study Group (1984) *Geography and Gender: An Introduction to Feminist Geography*, London: Hutchinson.

—— (1997) *Feminist Geographies: Explorations in Diversity and Difference*, Harlow: Addison Wesley Longman.

Woodward, D., Green, E. and Hebron, S. (1988) 'Research note: the Sheffield study of gender and leisure: its methodological approach', *Leisure Studies* 7, 1: 95–101.

Woodward, R. (1993) 'One place, two stories: the two interpretations of Spitalfields in the debate over its redevelopment', in G. Kearns and C. Philo (eds) *Selling Places: The City as Cultural Capital, Past and Present*, Oxford: Pergamon Press.

Wright, P. (1985) *On Living in an Old Country: the National Past in Contemporary Britain*, London: Verso.

Wynne, D. (1992) *The Culture Industry: The Arts and Urban Regeneration*, Aldershot: Avebury.

Yokeno, N. (1974) 'The general equilibrium system of space economics for tourism', *Reports for the Japan Academic Society of Tourism* 8: 38–44.

York, P. (1984) *Modern Times*, London: Heinemann.

Yuval Davis, N. and Anthias, F. (1989) *Women, Nation-State*, London: Macmillan.
Zukin, S. (1982) *Loft Living: Culture and Capital in Urban Change*, Baltimore and London: The Johns Hopkins University Press.

Index

Milton Keynes UK
Ingram Content Group UK Ltd.
UKHW020030071024
449327UK00032B/3003

9 780415 271660